The Ancient Murrelet

For all those who have sought the murrelet, in starlight, or in rain

Part of the royalties from this book go to the Laskeek Bay Conservation Society, to continue the work which we began

The Ancient Murrelet:
A NATURAL HISTORY IN THE
QUEEN CHARLOTTE ISLANDS

by
ANTHONY J. GASTON
of the
CANADIAN WILDLIFE SERVICE

illustrated by
Ian Jones

T & A D POYSER
London

© *Anthony J. Gaston*

First published in 1992 by T & A D Poyser Ltd
24–28 Oval Road, London NW1 7DX

United States Edition published by
ACADEMIC PRESS INC.
San Diego, CA 92101

All rights reserved. No part of this book may be reproduced, stored in a retrieval system, or transmitted in any form or by any means, electronic, mechanical, photocopying or otherwise, without the permission of the publisher

Text set in Linotron Baskerville
by Phoenix Photosetting, Chatham, Kent
Printed and bound in Great Britain by
Mackays of Chatham PLC, Chatham, Kent

A catalogue record for this book
is available from the British Library

ISBN 0 85661 070 4

Contents

Acknowledgements	xvii
Chapter 1 Introduction	1
PART I A GLOBAL GLIMPSE OF THE ANCIENT MURRELET	5
Chapter 2 A History of the Ancient Murrelet	7
What they look like	7
Who found the Ancient Murrelet?	9
. . . How it got its name	12
The genus *Synthliboramphus*	13
. . . And other relatives	14
How they appeared to early ornithologists	14
Discovering the distribution of the Ancient Murrelet	15
. . . And how they live	17
Conclusions	19
Chapter 3 Distribution, Status and Breeding Habitat	20
Distribution and status in Asia	22
The Aleutian Islands and the Bering Sea	25
The Alaskan Peninsula and the Gulf of Alaska	29
The Queen Charlotte Islands	30
Other areas south of Alaska	37
Breeding habitat	37
The other *Synthliboramphus* murrelets	39
Summary	42
Chapter 4 The Ancient Murrelet's Year	43
Chapter 5 The Ancient Murrelet at Sea	52
Diet	52
Morphology and feeding adaptations	55
Feeding behaviour	57
Non-breeding range	59
Marine habitat	62
PART II STUDIES AT REEF ISLAND	65
Chapter 6 Introduction	67
The Pacific Northwest	67
Geography of the Queen Charlotte Islands	69
Marine environments of the Queen Charlotte Islands	71

vi Contents

Terrestrial environments	74
Recent ecological changes in the islands	78
Role of the Ancient Murrelet in the ecology of the Queen Charlotte Islands	82
Diversity, stability and productivity	83

Chapter 7 The Island and the Work	**85**
How the study developed	85
The geography and vegetation of Reef Island	87
The fauna of Reef Island	89
The birds of Reef Island	91
. . . And adjacent waters	92
Methods	95
Measurements	101

Chapter 8 Attendance and Behaviour at the Colony	**104**
Voice	104
Pre-laying	107
Egg-laying	115
Incubation	117
Departure of family parties	119
How the chicks find their way to the sea	124
. . . And how they find their parents again	126
The first days at sea	128
Fledging and independence of the chicks	131

Chapter 9 The Behaviour of Non-Breeders	**132**
Timing of attendance	133
Singing and calling	134
Prospecting of burrows	136
What we think the non-breeders are doing	140

Chapter 10 Behaviour on the Gathering Ground	**142**
The location of the gathering grounds	142
What birds attend the gathering ground?	143
Behaviour on the gathering ground	150
What are the gathering aggregations for?	151

Chapter 11 Breeding Habitat and Burrows	**154**
The nest-site	154
The nest-chamber and cup	157
Measures taken to conceal the burrow entrance	158
Population on Reef Island	159
Variation in burrow density	161
Burrow occupancy	166
Continuity of occupation	167
Conclusions	169

Chapter 12 Eggs and Chicks	170
The eggs: size, colour and composition	170
Clutch size and the interval between layings	173
Egg recognition	175
Incubation and hatching	176
The effects of chilling on hatching and incubation	179
Time spent in the burrow by the chicks	180
Some adaptations of the chicks	180
Chapter 13 Timing of Breeding and its Effects	186
Timing and spread of laying	187
Year-to-year variation	189
Possible causes of inter-year variation	192
Individual variation	193
Summary	196
Chapter 14 Population Dynamics	198
At what age do birds begin to visit the colony, and how old are they when they begin to breed?	199
Movements within the colony	202
Annual survival	202
Reproductive success	205
Predation by deer mice	206
Brood size after leaving the colony	207
Age structure of the population	208
Movements among colonies	209
Summary	210
Chapter 15 Why Ancient Murrelet Chicks are Precocial	211
Chick rearing in seabirds	212
Chick departure strategies in the auks and their relatives	212
Previous explanations for chick departure strategies in auks	213
How do these ideas fit the evidence for Ancient Murrelets?	215
Some speculations on the origins of precociality	218
Chapter 16 Conservation	221
Some thoughts on conservation biology	221
Conservation of Ancient Murrelets	222
Epilogue	225
Rhyme of the Ancient Murrelet	226
Appendix 1 *Scientific names of animals and plants mentioned in the text*	227
Appendix 2 *Census details for Reef Island, 1989*	229
References	230
Index	245

List of Figures

3.1	*Breeding range of the Ancient Murrelet*	21
3.2	*Natural range of spruce* Picea *spp. and sockeye salmon* Oncorhynchus nerka, *and the major North Pacific current systems*	23
3.3	*Breeding sites of Ancient Murrelets in the Aleutian Islands and the Gulf of Alaska*	28
3.4	*Breeding sites of the Ancient Murrelet in the Queen Charlotte Islands*	31
3.5	*Changes in the distribution of breeding Ancient Murrelets on Langara Island since the 1950s*	33
3.6	*Breeding distribution of the Japanese Murrelet*	40
4.1	*Variation in the timing of breeding of Ancient Murrelets in different parts of their range*	47
4.2	*The relationship between the timing of egg-laying in Ancient Murrelets and sea-surface temperatures in the breeding area in April*	48
4.3	*The timing of different events in the Ancient Murrelet's breeding cycle at Reef Island, British Columbia*	49
5.1	*Diet of breeding and non-breeding Ancient Murrelets collected off Langara Island during March-June*	53
5.2	*Changes in weight, and numbers of euphausids per stomach, of Ancient Murrelets collected off Victoria*	55
5.3	*Winter distribution of the Ancient Murrelet*	60
6.1	*Map of the Queen Charlotte Islands*	70
6.2	*The food web involving Ancient Murrelets in the Reef Island area*	75
7.1	*Map of Reef Island, showing the position of the Ancient Murrelet breeding area, and our census transects*	88
7.2	*Temperature and rainfall records for Sandspit, British Columbia, during the periods of our study on Reef Island*	90
7.3	*Relative abundance of marine birds seen commonly in the vicinity of Reef Island*	95
7.4	*Proportion of birds trapped with brood-patches in relation to time of year*	100
7.5	*Position of chick trapping funnels used on Reef Island*	101
8.1	*Examples of* chirrup *calls, taken from songs recorded at Reef Island*	105
8.2	*Examples of* chatter *calling recorded during the first 2.5 s of greetings interactions, recorded during incubation at four different burrows on Reef Island*	106
8.3	*Examples of Ancient Murrelet songs recorded at Reef Island*	107
8.4	*Examples of other forms of Ancient Murrelet vocalizations recorded at Reef Island*	108
8.5	*Numbers of birds landing and chirrup calls recorded during the first 2 h of arrivals*	110

x List of Figures

8.6	Numbers of burrows entered at two plots on Reef Island in relation to the wind speed at midnight	111
8.7	The timing of first arrivals at Reef Island, in relation to the time of year	112
8.8	Number of nights that burrows were entered in relation to whether or not they were used for breeding	114
8.9	The number of egg-laying females trapped, in relation to the time at which other birds were captured	115
8.10	Numbers of eggs laid, in relation to the proportion of burrows entered that night	116
8.11	Visits to burrows between the laying of the first and second egg, in comparison with visits to burrows where eggs had not been laid	117
8.12	Duration of incubation shifts	118
8.13	Timing of first captures of chicks at the trapping funnels, in relation to time of year	120
8.14	Apparatus used to test the ability of chicks to use different cues in finding their way to the sea	124
8.15	Positions of Ancient Murrelet broods located at sea after their departure from Reef Island	130
9.1	Numbers of birds captured at different times of the season that had been actively excavating burrows	135
9.2	The proportion of breeders among birds captured at Reef Island, in relation to time after the start of arrivals	136
9.3	The proportion of burrows entered per night in relation to whether or not they had been used that year	137
9.4	The frequency with which unoccupied burrows were entered in the pre-laying, incubation and chick-departure periods	139
10.1	Position of Ancient Murrelet gathering grounds in the Queen Charlotte Islands	144
10.2	Results of gathering ground surveys carried out in the area of Reef and Limestone islands	145
10.3	Major gathering ground concentrations recorded around Reef and Limestone islands	146
10.4	The relationship between evening counts of birds flying over the Reef Island gathering ground, and the proportion of burrows entered that night	147
10.5	Corrected gathering ground counts made from Reef Island	148
10.6	Build up of gathering ground counts at Reef Island, in relation to the time before first arrivals at the colony	149
11.1	The location of burrows found on Reef Island	155
11.2	Entrance diameters of Ancient Murrelet burrows at Reef Island	155
11.3	The length of Ancient Murrelet burrows on Reef Island	156
11.4	Shapes of Ancient Murrelet burrows at Reef Island	156
11.5	Map showing the distribution of burrows found during the 1989 census of Reef Island	160
11.6	Burrows found per plot, in relation to slope	162

List of Figures xi

11.7	*Mean burrow densities, in relation to distance from the shore*	163
11.8	*Frequency with which burrows at Reef Island were occupied during 1985–89*	164
11.9	*Proportion of burrows occupied in successive years, in relation to whether or not breeding was successful in the first year*	168
12.1	*Fresh egg weight in relation to adult body weight in auks*	171
12.2	*Egg density indices in relation to the length of time for which the egg had been incubated*	176
12.3	*Numbers of days of neglect observed after the start of incubation at burrows checked only with temperature probes (hence undisturbed)*	178
12.4	*Number of days that chicks were present in their burrows before departure*	181
12.5	*Weights of chicks removed from burrows on the first day after hatching, compared with the weights of chicks trapped at the shore during departure from the colony*	182
12.6	*Weight losses of chicks reweighed at different intervals after the first day in the burrow*	184
13.1	*Dates of chick departures at Reef Island, based on captures in trapping funnels*	188
13.2	*Variation in different features of reproduction, in relation to the median date of clutch completion*	190
13.3	*The regression of egg size on date of laying in five years at Reef Island*	191
13.4	*Weights of chicks at departure, in relation to date*	192
13.5	*Median dates of clutch completion at Reef Island, in relation to mean minimum March temperatures*	193
13.6	*Dates of capture of adults with chicks, in different years, in relation to the median departure dates for each year*	194
13.7	*Changes in the date at which individual birds were caught with chicks in relation to the number of years that had elapsed between first and second captures*	195
14.1	*The date of trapping of second and third year birds, and of all birds without brood patches, at Reef Island*	201
14.2	*Proportions of Ancient Murrelets retrapped at Reef Island, in relation to the length of time elapsed since their first capture*	204
15.1	*Mean weights of meals delivered to nestling auks, in relation to adult weights*	214

List of Plates

Sunset over Louise Island, from Reef Island	69
Evening in the Misty Isles	71
Steller's Sea Lions on the rocks to the southeast of Reef Island	73
Inside the mature forest, Reef Island	76
An area of dense regenerating spruce on the site of a former windthrow	77
The research station on Reef Island, 1985. It has since been demolished	86
"Cassin's Castle" from the sea. Cassin's Auklet burrows were very dense on top of the small bluff in the centre	93
Sooty Shearwaters at dusk close to Reef Island (Photograph by Ian L. Jones)	94
Seawatching from the lookout near camp; Louise Island in the background (Photograph by Ian L. Jones)	96
Plastic tags placed in the mouth of a burrow	97
An Ancient Murrelet with a radio transmitter glued to the feathers of the lower back	98
A fully developed brood-patch on an Ancient Murrelet	99
One of the chick-trapping funnels	102
A brood of day-old Ancient Murrelet chicks removed from a burrow at Reef Island for ringing	127
Adult Ancient Murrelet on the forest floor (Photograph by Ian L. Jones)	133
Groups of Ancient Murrelets on the Reef Island staging area. Low Island is behind and to the right	143
Ancient Murrelets displaying on the staging area off Limestone Island (Photograph by Colin French)	150
Yves Turcotte weighing a murrelet's egg, at the mouth of a burrow	157
Plot D, Reef Island; prime nesting habitat for Ancient Murrelets	165
An Ancient Murrelet clutch, Reef Island	172
Section through the middle of a hard boiled Ancient Murrelet's egg, showing the very large size of the yolk	174
A clutch of eggs probably resulting from the laying of two females	174

List of Tables

3.1	*Estimates of Ancient Murrelet populations breeding in Alaska*	26
3.2	*Breeding populations of Ancient Murrelets in the Queen Charlotte Islands*	32
4.1	*Evidence for timing of breeding in the Ancient Murrelet and other Synthliboramphus species*	45
7.1	*Brood-patch development of females trapped with oviduct eggs*	99
7.2	*Measurements of Ancient Murrelets at Reef Island*	102
8.1	*Lengths of incubation shifts at Reef Island, and the proportion of incubation time spent in shifts of different lengths*	118
10.1	*Counts of Ancient Murrelets on boat transects of the gathering grounds adjacent to Reef and Limestone islands in 1989*	147
11.1	*Estimate of numbers of Ancient Murrelet burrows on Reef Island, 1989*	161
11.2	*Ancient Murrelet burrow dimensions at Reef Island*	167
12.1	*Egg mass in relation to female body mass*	170
12.2	*Dimensions and mass of first and second laid eggs measured at Reef Island in 1984 and 1985*	171
12.3	*Egg size in Ancient Murrelets*	173
12.4	*Incubation periods and number of days of incubation*	177
13.1	*Median dates of departure of chicks captured in three different funnels*	189
13.2	*Correlation coefficients for the date of capture of individual breeders in burrows with chicks in different years*	195
13.3	*Correlations between the volume index of eggs laid in the same burrow in different years, where at least one member of the pair was the same*	196
14.1	*Numbers of chicks banded and numbers retrapped in subsequent years*	199
14.2	*Weights and wing-lengths of birds captured at Reef Island*	200
14.3	*Years of banding of birds retrapped in subsequent years*	203
14.4	*Regression estimates of adult annual survival*	204
14.5	*SURGE estimates of adult survival and recapture probabilities*	205
14.6	*Reproductive success of Ancient Murrelets at Reef Island*	206

Acknowledgements

I was first introduced to the Ancient Murrelet, and to Reef Island, by Gary Kaiser and his 1983 field crew, which was led by Moira Lemon and Michael Rodway. These three people were subsequently the most influential in helping me to set up and carry out the six year study on Reef Island which forms the mainstay of this book. Their contributions ranged from advice on handling inflatable boats, and arranging for marine telephones, to providing sofas for transient field crews and servicing outboard motors. In addition, they all read and commented on the manuscript. Their support, and the backing of the Pacific and Yukon Region of the Canadian Wildlife Service, was invaluable. I am deeply indebted to them.

During the work on Reef Island, I was fortunate in having the assistance of many dedicated people, who provided me with insights that I might never have obtained on my own. I would like especially to thank Ian Jones, whose thesis and publications form the basis for much of Chapters 8 and 9, and who supplied the sonograms and the excellent line drawings which sprinkle the book. Ian also provided sage advice on the whole text. Steven Smith, David Noble, David Powell and Andrea Lawrence were all involved for more than one season, put their individual stamp on the work, assisted me in damaging the local fish stocks, and commented on parts of the manuscript. David Duncan, who carried out post-doctoral research on the energy reserves of nestling murrelets, contributed many new ideas. For assistance with field work, I also thank Yves Turcotte, Lawrence Turney and Bruce Falls (1984), Paul Jones and Chris Risley (1985), Louise White and Pierre Mineau (1987), Alan Burger, Michael Fulmer and Keith Hobson (1988), and Jane Witney, Susan Johnson, Timothy Lash, Jean-Louis Martin and Kathy Heise (1989).

The unfailing kindness that is typical of people in the Queen Charlotte Islands was extended to me on many occasions. I must mention especially the assistance of Keith Rowsell and family, and Terry and Charlotte Husband and family. Lilian Brown supplied me with reminiscences of the Haida traditions of harvesting Ancient Murrelets. The local representatives of the British Columbia provincial wildlife branch were another source of support, and, in particular, I would like to thank Keith Moore, Lorne MacIntosh and Al Edie.

Information supplied by fellow ornithologists on the distribution and biology of Ancient Murrelets is mainly acknowledged in the text. However, I must mention Spencer Sealy and Harry Carter, who kindly supplied me with unpublished data on the winter diet, Ken Summers, who provided details of his previous work on Reef and Limestone islands, John Piatt, Alan Springer and Doug Forsell, who provided unpublished information on behaviour and diet in Alaska, and Ed Bailey, who corrected the section on Alaskan populations. Cal Lensink allowed me access to his personal library, Natalya Litvinenko kindly supplied pre-publication material on Ancient Murrelets in the Soviet Union, and Hito Higuchi corresponded on the state of the Japanese Murrelet in Japan. Wayne Campbell, of the Royal B.C. Museum, provided me with access to unpublished survey data filed at the museum, and also to the specimen collection. I offer special thanks to Vernon Byrd, of the U.S. Fish and Wildlife Service, who kindly took on the onerous job of criticizing the final manuscript.

In dealing with literature in Russian, I was much assisted by Michael Wilson, of *Birds of the Western Palaearctic*, and in searching the older references, Linda Birch, of the Alexander Library, Oxford, was very helpful. Theresa Morgan kindly translated descriptions in Latin. For lengthy translations from Russian and Japanese, I am indebted to the translation services of the Secretary of State's Office, Government of Canada. Many thanks, also, to the drafting unit of the Department of Environment, for the splendid text figures. Finally, I would like to thank my immediate supervisors, Hugh Boyd, and Tony Keith, and my wife, Anne-Marie, for their support and encouragement during the work on Reef Island, and during the writing of this book.

CHAPTER 1

Introduction

In Farid Ud-din Attar's 12th century Sufi poem, *The conference of the birds*, the Hoopoe persuades the other birds to undertake a quest in search of their king. The Heron protests that he is too fond of the sea to undertake the journey which will cross deserts and mountains. The Hoopoe replies

> ... You do not know
> the nature of this sea you love: below
> its surface linger sharks; tempests appear,
> then sudden calms—its course is never clear,
> but turbid, varying, in constant stress;
> Its waters taste is salty bitterness ...
> The diver plunges and in fear of death
> Must struggle to conserve his scanty breath;
> The failure is cast up, a broken straw.
> Who trusts the sea? Lawlessness is her law.
> [Attar, trans. Darbandi and Davis, 1984.]

The sage advice of the Hoopoe has not prevented many birds from being drawn to a marine way of life. The sea was the home of some of earth's earliest birds. Like the reptiles that had preceded them and the mammals that followed them in the colonization of the marine environment, these birds were confronted with the challenge of maintaining an air-breathing respiratory system, not only for their adult stages, but for their earlier developmental stages, as embryos. Having never evolved viviparity, birds were faced with the need to lay their eggs where they could respire through an air-breathing shell, and that meant returning periodically to land.

2 Introduction

The unbroken link to their original terrestrial environment has moulded many features of the biology of marine birds. Practically all have adopted the strategy of rearing their young on land to more or less full size, while continuing to find food for themselves and their offspring at sea. The consequent commuting between nest and feeding area has been a crucial factor in the evolution of many features of seabird biology, the most conspicuous of which is the small clutch size, frequently only a single egg (for more extensive treatments, see Ashmole, 1963, 1971; Lack, 1968; Goodman, 1974; Ricklefs, 1983; Pennycuik et al., 1984).

Out of approximately 70 extant genera of seabirds, only the genus *Synthliboramphus*, to which the Ancient Murrelet belongs, has adopted the alternative strategy of producing young that are not fed at all in the nest site but accompany their parents to the feeding area soon after hatching. In fact, if this retiring and little-known genus had succumbed to the eventual fate of all genera by going extinct before it was discovered by science in the 18th century, we might never have considered such a development pattern to be feasible for marine birds. Their existence poses a challenge of evolutionary interpretation; how can we explain the evolution of such behaviour, and its associated physiological adaptations, in this genus without jeopardizing our belief that the strategies of other seabirds are well-fitted to their particular ecology? Has evolution in most seabird lineages somehow overlooked an important alternative pathway, or is the ecology of the *Synthliboramphus* murrelets so special that it has led to the evolution of a unique strategy which would not be viable for any other seabird? I hope that, in describing the biology of Ancient Murrelets *Synthliboramphus antiquus* in the Queen Charlotte Islands of British Columbia, I shall shed sufficient light to be able to provide at least a tentative answer to the paradox posed by their precocity.

Although Ancient Murrelets breed throughout the temperate north Pacific, I have concentrated my attention on the species in the Queen Charlotte Islands, where I conducted six seasons of intensive breeding biology studies on the species (1984–89). This is also the area where the only other intensive study of the species has been carried out, by Spencer Sealy (Sealy, 1972, 1975a, b, 1976) in 1970–71. Moreover, during the period of my research, my colleagues in the Canadian Wildlife Service, Gary Kaiser and Kees Vermeer, and a field-crew ably led by Moira Lemon and Michael Rodway, surveyed the whole of the Queen Charlotte Islands to record the position and size of Ancient Murrelet colonies (and those of other seabirds). In the process they recorded much valuable additional information on their ecology and behaviour (Rodway et al., 1988; Rodway, 1991). In contrast, there has been no intensive study of Ancient Murrelets in the rest of their global range, and indeed not much is known about them elsewhere, apart from anecdotal observations or rough estimates of colony size.

My research in the Queen Charlotte Islands was very much a team effort. I was very lucky to have the assistance and fellowship of many outstanding individuals. I shared with them, over many hours of discussion around the cabin stove, over meals, or in the darkness of the forest as we awaited the arrival of the murrelets, many ideas about the murrelets' biology. In recognition of the indispensable role played by my companions, I have used the first person plural in describing most of our activities and our thoughts about the birds. Only when I come to set forward ideas that I have developed subsequently, or that may be controversial, or where I am specifically describing my own experience, do I use the singular.

One other piece of wording that may strike the reader is the use of North American names for the birds. I felt that this was appropriate for several reasons. The book involves a bird which, in the English speaking world, occurs only in North America, and practically all of the studies that I report have been undertaken there. Moreover, in referring to the other auks, the use of the English "guillemot" is ambiguous, because it can refer to members of either of the genera *Uria* and *Cepphus*, whereas in North America "guillemot" applies only to *Cepphus*. Use of the North American names has saved me some circumlocution. For my English readers, the Common Murre is the Common Guillemot, and the Thick-billed Murre is the Brunnich's Guillemot. I hope that my erstwhile compatriots will forgive me this trespass.

Over the past decade, there has been a great increase in interest in the Queen Charlotte Islands among ecologists, naturalists and lovers of wild places. The combination of magnificent old-growth forests, luxuriant marine life and endemic terrestrial fauna found in these islands, makes them outstanding, even in British Columbia, which remains a naturalist's paradise. The Ancient Murrelets, with their unusual biology and their attachment to some of the most visually spectacular parts of the archipelago, are one of the jewels in the islands' crown.

Despite their relative isolation, things are beginning to change very rapidly in the Queen Charlotte Islands. Although a recent decision to turn the southern part of the archipelago into a national park means that natural environments will be safeguarded against developments on land, developments at sea continue to pose a potential hazard. New fisheries are being explored and may lead to further drastic alterations of marine food webs. Oil drilling and sea-floor mining in the adjacent waters of Hecate Strait are a possibility, in the long-run perhaps a certainty. At the same time the number of people visiting the islands as tourists is increasing rapidly, and is bound to have some impact on birds as susceptible to disturbance as Ancient Murrelets. There seems little chance that even such remote and retiring birds as the murrelets will escape indefinitely the changes that have engulfed the rest of the globe. It therefore seems appropriate to record their biology and ecology as it exists now, so that a firm baseline is available against which future investigators can measure the effects of change, whether it be fisheries, oil development, tourism, pollution, or the increase in global temperatures that climatologists are currently forecasting.

Part I
A Global Glimpse of the Ancient Murrelet

CHAPTER 2

A history of the Ancient Murrelet

What they look like; how the Ancient Murrelet got its name; other species of the genus and their position among the auks; a short history of how knowledge about the Ancient Murrelet's distribution, and biology was accumulated.

WHAT THEY LOOK LIKE

The Ancient Murrelet is a small marine bird, weighing between 170 and 270 g, belonging to the family Alcidae, the auks. Like other members of its family, it is well adapted to swimming underwater, having an elongated, torpedo-shaped body, with short legs, set far back, and short but strong wings, which it uses as paddles to propel itself while submerged as well as for flying. The auks are sometimes thought of as the Northern Hemisphere counterparts of the penguins, which they resemble in diving deeply and swimming with their wings. In fact, the original *penguin* of the French was the Great Auk, a large flightless member of the family, which is now extinct.

In summer, the adult Ancient Murrelet is largely dove grey above and white below. The head and throat are entirely black, except for a "tonsure" of long, floppy, white feathers which surrounds the crown and is somewhat elongated at the back of the neck. Where the black of the head meets the grey of the back and mantle, there is a zone of pepper-and-salt speckling, created by the presence of a variable number of elongated silvery feathers known as filoplumes. The bill is pale yellow, with a variable amount of black along the top of the upper mandible and at the base, while the tarsi and feet are pale blue, with black claws. There is no difference in appearance between the sexes, nor is there any difference in measurements, except that females have slightly shallower bills than males and slightly narrower heads (Chapter 6). Females also have slightly, though not significantly, longer wings. In the closely related Craveri's Murrelet, females do have significantly longer wings than males (Van Rossem, 1926; Jehl and Bond, 1975).

The mantle and upper wing-coverts of the Ancient Murrelet are grey, with the primaries blackish-brown, while the underwing-coverts are white. In the air, Ancient Murrelets, like all auks, flap vigorously at all times, but their wing-beats are comparatively shallow, with wings held rather stiff, and they tend to bank from side to side, so that their white underparts flicker as they zoom along just above the waves.

In the autumn, adult Ancient Murrelets moult completely, adopting a winter plumage. Very little is known of the moult, which presumably takes place far out at sea. Birds collected in November have very fresh plumage, so the moult is completed between the end of breeding, in July–August (depending on the locality) and the end of October (Flint and Golovkin, 1990; pers. obs.). The winter plumage is similar to the summer plumage, except that the throat is white, with a variable amount of sooty smudging on the chin, and the tonsure of white feathers is much reduced, becoming only an obscure flecking. The filoplumes largely, or completely, disappear, and the mantle appears somewhat more sooty grey than in summer. At a distance, in winter, the birds can appear blackish above, giving them some resemblance to Marbled Murrelets at the same season.

A second moult, involving only the body feathers, occurs from November onwards, and transforms the birds back to their breeding plumage. About one third of birds in the British Columbia Provincial Museum, collected in November, have black throats, although none have fully developed tonsures. By January the proportion with black throats has risen to about two-thirds, and by March all birds trapped at the breeding colonies have black throats and fully developed tonsures, although a few show small white flecks on the throat.

Young birds in their first plumage resemble winter plumage adults, except that the bill is usually somewhat thinner (Flint and Golovkin, 1990) and paler, and there is no sign of black on the throat (pers. obs.). They undergo a post-juvenile moult in October into normal winter plumage, but according to Kozlova (1957), there is no spring moult in the first year, so birds in their first summer are in worn winter plumage. However, two first year birds trapped at Reef Island in May were in summer plumage. Moreover, we never saw any birds in winter plumage in Hecate Strait in summer, so if most first year birds do remain in winter plumage they presumably remain well away from the breeding colonies. By late winter, birds in their first year are hard to distinguish from adults. As no birds of known age have been collected at that stage, it is not certain how they differ from older birds. However, we presume that those birds collected in February which still maintain winter plumage, and show no development of the tonsure, or filoplumes, are probably in their first year. Birds trapped at the colony in their second summer had shorter wings than older birds (Chapter 12), so this is probably true of first-year birds as well, but no information is available.

Most Ancient Murrelets are considered to constitute a single subspecies, *Synthliboramphus antiquus antiquus*, but another, *S. a. microrhynchos*, has been described from the Commander Islands by Spepanyan (Flint and Golovkin, 1990). The birds in the Commanders apparently have less white speckling around the neck, and smaller bills, than the nominate race. Specimens of *S. antiquus* that I examined in the British Museum (Natural History), the Royal British Columbia Provincial Museum and the American Museum of Natural History, which originated in areas from Japan to British Columbia, showed little plumage variation. Compared with certain storm-petrels, Ancient Murrelets seem to be fairly uniform in their appearance.

WHO FOUND THE ANCIENT MURRELET?

The naming of most birds is a relatively straightforward process. Someone collects a bird that they think is new to science and then they, or someone else, describes it in a published document, designates a type specimen, and gives the locality where it was discovered. However, for some birds, the process of acquiring a name is less clearly documented. The Ancient Murrelet is one of these. To unravel how it got its name we need to do a little detective work.

There is some uncertainty about the exact moment at which the Ancient Murrelet came to the attention of science. The first formal name, *Alca antiqua*, was attached by Gmelin (1789) in his edition of Linneus' *Systema Naturae*. He based it, not on his own observations, but on the description given by Thomas Pennant in his *Arctic Zoology* (1784: 512; Pennant's book actually spelled the name, "Antient Auk", an obsolete rendition of "ancient", altered by Latham, 1785). Pennant, in turn, referred to a manuscript of Pallas. However, when the latter published his great *Zoographia Rosso-Asiatica*, in 1811, he gave primacy to the description of Steller, a young German whom he had hired to investigate the natural history of the easternmost parts of the Russian empire (Stejneger, 1936; Ford, 1967). Pallas used the name *Uria senicula*. To complicate the matter further, it appears that Steller's description reached Pallas only through Krascheninnicof's *History of Kamchatka*. The latter book I failed to trace, but some extracts from it are given by Pearse (1968), who relates that Steller, as Krascheninnicof's senior, insisted that any notes or specimens that Kraschenninicof took be handed over to him. Consequently, it is very hard to tell exactly who saw what, or where they did so.

My conclusion is that the bird was probably first described by Steller, although some of his information may have derived from Krascheninnicof. Steller saw it in 1741, while taking part in Bering's second voyage across the North Pacific.

On 5 September, off the Shumagin Islands (just south of the Alaska Peninsula) Steller recorded; "a very beautiful black-and-white pied diver, never before seen". Stejneger (1936) and Ford (1967) both interpreted this to refer to the Ancient Murrelet, which by that season might have been in black-and-white, nonbreeding plumage. However, on 28 October, Steller reported

> A small species of diver known as "*Starik*" [this is the Russian name for the Ancient Murrelet] flew on board our vessel during the night; these birds habitually pass the night on the rocks and fly against everything they see only dimly near at hand, like owls in the daytime. Because of this, large numbers of them around Avatcha are caught alive with the hands by merely sitting down near them covered with a mantle or *kuklanka* (a fur coat of the Kamchadals), under which they gather, as if in a ready nest. [Pearse, 1968.]

This description, and the Russian name, suggest that Steller already knew about the bird, perhaps from the accounts of local people. Consequently, I am inclined to think that the "pied diver" met with on 5 September must have been something else, perhaps a winter-plumage Marbled Murrelet, not discovered by science until Cook's last expedition, in 1778 (Stresemann, 1949). Moreover, the first bird noted (on 4 June) in Steller's journal of the voyage occurs in the following passage, concerning an island not far from the expedition's starting point, in Avatcha Bay:

> An insular mountain of stone called Starischkow, after the name of the bird that frequents it in great abundance. About the size of a pigeon, with bluish bill and small feathers, of a bristling kind; head of purple colour, having a circle of white feathers in the middle, which are thinner and longer than the rest. Its neck is black above with white spots underneath. Body white, short, large feathers of wings blackish and the rest of blue, its sides and tail black; its feet red and trebly indigitated [*sic*] with a web between each and its claws black. On little islands of Kamchatka they are found in vast numbers and caught by the Kamschadals also with great facility. [Pearse, 1968.]

This description fits the Ancient Murrelet in some respects, but is very inaccurate in others, especially the red feet (actually bluish), which is perhaps why Stejneger thought it referred to the Parrakeet Auklet. Actually, none of the auklets have red feet. Only puffins and Pigeon Guillemots fit this part of the description. The confusion suggests that Steller, a very meticulous observer, was simply going on hearsay at that point. The same account, almost word for word, appears in Krascheninnicof's book (Pearse, 1968) and he names the birds involved (*starik*), adding Steller's account of the murrelets running up the sleeves of cloaks laid on the ground, a note also included by Pallas. The latter confirmed that the name of the island derived from the Russian name of the Ancient Murrelet. More surprisingly, perhaps, exactly the same account of Starichkowa and its birds appears in King's journal of the last Cook expedition, after the death of Cook (Pearse, 1968). Starichkova Island (as it is now transliterated) still supports several thousand pairs of breeding Ancient Murrelets (Chapter 3). It seems almost certain that the Ancient Murrelet was the bird referred to in Steller's note of 4 June.

Steller's discovery of the Ancient Murrelet is one of the lesser known feats of that remarkable naturalist, probably obscured by the greater accessibility of Pennant's work. However, science still gives priority to Pennant for the first published description, which was brief but clearly recognizable:

> With a black bill, crown and throat: on each side of the head a short whitish crest: on the hind part of the neck are numbers of white, long, loose and very narrow feathers, which give it an aged look: wings, back and tail sooty: breast and belly white. Size of the former [Little Auk]. Inhabits the west of North America to Kamtschatka [*sic*] and the Kurile Islands.

There is no other bird which remotely agrees with this description. It is definitely an improvement on those given by Steller. Pennant presumably saw the specimen brought back by the last Cook expedition (Stresemann, 1949).

Such was the speed of communication among scientists at that time that Latham's (1785) *General Synopsis of Birds* was already quoting Pennant's work of the previous year, and even speculating a little about the long white feathers at the sides to the neck, which "may *perhaps* [his italics] be erected at the will of the bird, as a ruff". Nice try, but actually they do not use them in this way, although the hunched pose adopted during certain displays given at sea does accentuate the filamentous white feathers.

The fact that the range given by Pallas included both the Asian and American shores of the Pacific, suggests that he knew of the specimen taken by the Cook expedition off Unalaska (Stresemann, 1949). Brandt (1869) asserted that Steller saw the species off, "Caput Eliae" (St Elias Head, Alaska), although on what authority is not clear. This would have made the Ancient Murrelet one of the first birds that Steller encountered in continental North America, along with the jay that bears his name (Steller's Jay), which was collected during his one brief foray on shore. In July, when the expedition made its American landfall, Ancient Murrelets would probably have been common along that part of the Alaskan coast. However, there is no record in Steller's surviving notes. The expedition members might well have encountered them also around Bering Island, in the Commander (Komandorskii) islands off Kamchatka, where they were eventually forced to winter.

Pallas included Penzhinskaya Gulf, at the head of the Sea of Okhotsk, within the Ancient Murrelet's range, presumably on the authority of Kraschenninicof. It is hard not to speculate that much of Steller's information on the Ancient Murrelet might have come from Kraschenninicof, who travelled widely in Kamchatka and the Kurils. All kinds of speculations are possible but, whatever actually occurred, it is clear that Steller left a coherent account of the Ancient Murrelet, which formed the basis for knowledge on the species' biology until late in the 19th century.

...HOW IT GOT ITS NAME

As with most species, it took some time for scientists to settle on a name acceptable to everyone. Gmelin (1789) used *Alca antiqua*, but the genus *Alca*, originally more-or-less equivalent to the modern family Alcidae, was much too big and diverse to be tolerated by later taxonomists. Brandt (1837) invented the tongue-twister *Synthliboramphus* (literally, "compressed beak"), to give us the present *S. antiquus*, designating *Alca antiqua* of Gmelin as the type. He defined it originally as a subgenus of *Brachyramphus*, but in his major work on auks (Brandt, 1869) he upgraded it to full generic status.

Inclusion in Gmelin's list ensured relative stability for the Ancient Murrelet's Latin name, although it managed to garner several other descriptions. Audubon, who included the species in his monumental *Birds of America* (1827–30), used *Mergulus antiquus* after Bonaparte (1838), and called it "Black-throated Guillemot". In fact, his plate includes two birds so named. One, clearly an Ancient Murrelet in summer plumage, is captioned "adult", while the other, equally clearly an adult Kittlitz's Murrelet in summer plumage, is captioned "young" (this error was identified by Brandt, 1869). Audubon's mistake is hardly remarkable, because he worked from skins and presumably relied on someone else's testimony as to the relatedness of the birds. More surprisingly, editions of *Birds of America* reprinted in 1937 and 1966 both failed to correct the mistake—some evidence of how unfamiliar most ornithologists were, even quite recently, with the small Pacific auks. By the time Audubon (1839) came to write his *Ornithological Biographies*, he reverted to the more common *Uria antiqua*.

The English appellation, "Ancient", agrees with the Russian name *starik* (old man), and refers to Pennant's, "white, long, loose and very narrow feathers... which give it an aged look". The English name also took some time to stabilize. Pennant's "Auk" was soon dropped, but Audubon's "Black-throated Guillemot" continued to be used until the late 19th century. Until the 1880s all the murrelets, as well as the present guillemots (*Cepphus*) and murres (*Uria*), were known as "guillemots", or "auks". Latterly, in North America, the term "guillemot" has been restricted to species of the genus *Cepphus*, the names murre and murrelet coming into use towards the end of the century. By the mid-1880s the name "Ancient Murrelet" was in use (Nelson, 1887; Coues 1890). Both used Black-throated Guillemot earlier in the decade. Coues (1890) switched "Temminck's Auk" to "Japanese Murrelet" at the same time. The name Ancient Murrelet became more or less fixed by its use in the first American Ornithologist's Union Checklist (AOU, 1886) and by Ridgeway (1896), in his *Manual of North American Birds*. Since 1900, only a small minority of authors have used any other English name. Why the specific designation "Black-throated" was dropped in favour of Pennant's original "Ancient" is not clear, except that the latter had the virtue of priority, and coincided with the Latin appellation.

Not everyone thought first of the plumage when naming the birds. In the Queen Charlotte Islands, the Haida Indians, who were very familiar with Ancient Murrelets and their habits, called them *"skinkana"*, which means "night pigeons". Indeed, the plump bodies and fluttering flight as they come in to land on the colony are somewhat reminiscent of pigeons. More importantly, from the perspective of the Haida, a plucked murrelet yields about as much meat as a dove. They were harvested at their colonies in large numbers until a few years ago.

THE GENUS *SYNTHLIBORAMPHUS*

Until recently, the only other species included in the genus *Synthliboramphus* was the Japanese, or Crested, Murrelet, *S. wumizusume*. The specific name was bestowed by Temminck (1850), and derives from the Japanese name for both Ancient and Japanese murrelets, meaning "sea sparrow". The designation is very appropriate, considering their sparrow-like calls, and stubby, somewhat sparrow-like bills. It has also been called Temminck's Guillemot, or Temminck's Murrelet. This bird is very similar to the Ancient Murrelet, except that it has a longer bill, less white on the sides to the neck, and in breeding plumage it possesses very long plumes at the back of the crown giving it a loose, floppy crest. This character has led to the Japanese name *kanmuri-umizusume*, the word *kanmuri* referring to the feathered head-dress of a chief. Both species breed in Japan, although the Ancient Murrelet has only a tenuous hold there (Chapter 2). The Japanese Murrelet breeds nowhere else.

Since 1983 (AOU, 1983) the genus *Synthliboramphus* has been expanded to include the two other precocial murrelets, Xantus' *S. hypoleucus* and Craveri's *S. craveri*. Until 1983 these two were placed in a genus of their own, *Endomychura*. They are very similar to one another in appearance and habits. There was some debate at one time about whether they constituted separate species (Van Rossem, 1926). This has been resolved by the discovery by Jehl and Bond (1975) that the two species overlap without interbreeding on the west coast of Baja California. However, the fact remains that the present *Synthliboramphus* contains two closely related pairs of species; on the one hand, *antiquus* and *wumizusume* with rather deep, compressed bills, and black throats and nuptial plumes in summer; on the other hand, *hypoleucos* and *craveri*, with slender, dagger-like bills, and no seasonal change of plumage. Hence the suppression of *Endomychura* has slightly obscured relationships within the group. The latest check-list of the American Ornithologists' Union calls the two species pairs, "superspecies" (AOU, 1983), which goes some way to re-asserting their affinities.

Because all four present members of *Synthliboramphus* have fully precocial chicks, I include some discussion of their biology within the scope of this book. The non-precocial murrelets (Marbled and Kittlitz's), both relatively poorly known because of their inaccessible breeding sites, are not dealt with.

...AND OTHER RELATIVES

The small Pacific alcids are divided into two groups on the basis of structure and ecology; the murrelets, which are relatively slender-billed and seem to be generalist feeders, particularly well adapted for underwater swimming (Kuroda, 1954, 1967) and the auklets, which have deeper, thicker bills and are less well adapted for underwater swimming but better adapted for moving about on land, and which specialize in plankton feeding (Bedard, 1969a). The Rhinoceros Auklet, it should be noted, is not an auklet at all, but is related to the puffins (Strauch, 1985).

The auklets, with the exception of Cassin's Auklet and the Whiskered Auklet, visit their colonies during daylight (Cassin's Auklet also does so occasionally, and Whiskered Auklets visit their colonies in daylight frequently during the pre-laying period). All, except Cassin's Auklet, possess brightly coloured beaks and ornamental plumes, which must function as visual signals. By contrast, all the murrelets visit their breeding sites at night (Marbled Murrelets may do so in daylight occasionally), their bills are relatively dull, their plumes are either less spectacular that those of the diurnal auklets (Ancient and Japanese Murrelets) or are absent altogether, and they concentrate on vocal communication (as do the nocturnal Cassin's and partially nocturnal Whiskered Auklets).

For some time the superficial similarity among the various species of murrelets led to all being placed in the same genus, *Brachyramphus*. Although Brandt (1869) removed the Ancient and Japanese Murrelets from this cluster fairly early, Craveri's and Xantus' Murrelets remained in *Brachyramphus* until given their own genus by Ogilvie-Grant (1898), who originally named the genus *Micruria*. *Endomychura* was coined by Oberholser (1899), because the former name was preoccupied. The first American Ornithologists' Union Checklist included them within *Brachyramphus*, but *Endomychura* was used consistently from the beginning of the 20th century (AOU, 1910). The classification was clarified along modern lines by Storer (1945), who divided the precocial murrelets from the non-precocial species on grounds of morphology and plumage differences. In doing so he reinstated an arrangement first suggested by Brandt (1869), who defined *Synthliboramphus* in the first place. A more recent, and more detailed, analysis by Strauch (1985) supports the combining of *Synthliboramphus* and *Endomychura*, and places them closest to the guillemots (*Cepphus*), these genera forming a tribe, Cepphini. In this classification the *Brachyramphus* murrelets form a separate tribe, although phyllogenetically closer to the Cepphini than to the other auks.

HOW THEY APPEARED TO EARLY ORNITHOLOGISTS

The descriptions of Pennant and Pallas may have been based on the same information, from Steller, but their descriptions differed somewhat. Pallas described the back as black (*niger*), in contrast to Pennant's "sooty". Gmelin, presumably taking his lead from Pennant, used *fuliginosa*. In fact the back is grey at all seasons, perhaps slightly duskier in winter, a fact that caused Coues (1868) to wonder whether more than one species was involved. None of the 18th century authors described the non-breeding plumage, giving a single description which included the "old man" plumes, which disappear in winter.

Audubon's illustration in the *Birds of America* remained for long the only picture of an Ancient Murrelet. It is a fine rendition, but the feet are shown as yellow. In fact, in life, the feet are bluish-grey to flesh-coloured on the tarsus and toes, darker on the webs. However, the feet of museum specimens generally look orange-brown, so they alter on drying out, and this was presumably the cause of Audubon's mistake. In the British Museum (Natural History) there is a specimen labelled "*Mergulus antiquis* [*sic*] Bonap. original of his description . . . original also of Mr Audubon's figure and description". The feet on this specimen are paler than those of any other in the collection, being almost straw-coloured. Sadly, the specimen is in poor shape, having lost many of the feathers from the head and neck. Audubon's understandable error over the colour of the feet was disseminated by later authors who took their descriptions from Audubon's plate (e.g. Brewer, 1840).

Brandt (1869) gave correct descriptions of both winter and summer plumage, as well as describing the down plumage of the chicks (after Vosnessenski, a reference that I have failed to trace). Also, by this date, Whitely (1867) had supplied the British Museum with several specimens in winter plumage, taken off Japan. From this point on, the identity of the species had been settled for all seasons.

DISCOVERING THE DISTRIBUTION OF THE ANCIENT MURRELET

Details of the Ancient Murrelet's distribution accumulated fairly rapidly after it was described, and by the mid-19th century Coues (1868) was able to identify its breeding range, more-or-less correctly, as extending across the northern Pacific from Japan to Washington State. By the late 19th century the British Museum (Natural History) boasted 24 specimens, from Korea, East Siberia, Japan (14), the Kuril Islands, Kamchatka, the Commander Islands, and northwestern North America (Ogilvie-Grant, 1898). Interestingly, none of these originated from Canada, and Ogilvie-Grant gave the range only as "Japan to Alaska", ignoring Coues. The presence of large numbers of Ancient Murrelets in British Columbia was not apparent from the ornithological literature until much later. In fact, the first record for British Columbia appears to be Fannin's (1891) check-list. Macoun and Macoun (1909), in their *Catalogue of Canadian Birds*, were still not certain that the species bred in Canada, stating merely that "Rev. J. H. Keen reports it rare in the Queen Charlotte Islands."

Breeding in the Queen Charlotte Islands was first reported by Green (1916, quoted by Drent and Guiget, 1961) and Hoffman (1924) reported an Ancient Murrelet's nest in Washington State in 1924. Brooks and Swarth (1925), in their list of birds in British Columbia, mentioned large numbers breeding on Langara Island, presumably deriving their information from Green. In the 1930s we have two published first-hand accounts of nesting in the Queen Charlotte Islands (Darcus, 1930; Cumming, 1931), the accounts again dealing with Langara Island, although Darcus mentioned that Indians reported Ancient Murrelets breeding on "Hippo" (presumably Hippa) Island, about 80 km south of Langara, on the west coast of Graham Island. Surprisingly, Taverner's (1934), *Birds of Canada*, for long the bible for Canadian ornithologists, maintained that there was no certain breeding record for the country.

Langara Island, off the northwest tip of the archipelago, was then a huge colony, and within sight (on a clear day!) of Forrester Island, where Willett (1914, 1915) and Heath (1915) had observed Ancient Murrelets to be extremely abundant. The colonies of the south Moresby area remained unknown to the outside world for another two decades. Munro and Cowan (1947), in their *Review of the Bird Fauna of British Columbia*, mentioned breeding only on Langara, Lucy (close to Langara) and Frederick islands, all off the northwest coast. It remained to Charles Guiget and his collaborators at the British Columbia provincial museum to initiate a thorough exploration of the fauna of the Queen Charlotte Islands from 1950 onwards. Most of the major colonies were recorded by 1972 (Guiget, 1972; Summers, 1974), but a few have been added in the past decade (Rodway *et al.*, 1988) and small groups of a few hundred nests may, even now, await discovery.

In the 1970s, the Provincial Museum, in the persons of Wayne Campbell and Michael Rodway, set out to catalogue, and roughly census, all the seabird colonies of British Columbia, carrying out extensive surveys covering all the islands (Campbell and Garrioch, 1979). From 1983 onwards Gary Kaiser, Kees Vermeer, and Moira Lemon of the Canadian Wildlife Service, again with Michael Rodway, extended this work to include intensive censuses of all the surface nesting seabirds and burrow nesting auks. The survey of the Queen Charlotte Islands was completed in 1987 (Rodway, 1988; Rodway *et al.*, 1988, 1991) and forms the basis for the distributional data presented in Chapter 3.

Flying in on the twice-daily jet service to Sandspit airport nowadays, it is hard to realize how isolated the Queen Charlotte Islands were, even 20 years ago. It was not that the area was unpopulated, or that the local people were unaware of their environment, but news from the islands emerged only fitfully, and most people who knew of the birds saw nothing extraordinary in their presence. This obscurity, combined with the retiring, nocturnal habits of the Ancient Murrelet, and other burrow nesting seabirds, made the islands of South Moresby the last major seabird breeding area in the affluent parts of the globe to be described.

...AND HOW THEY LIVE

The habits of Ancient Murrelets must have been well known to Aleuts, Haida and other native peoples resident in their breeding areas, because several early accounts of the bird mention harvesting at the colonies (Steller, in Pearse, 1968, also quoted by Pallas and Kraschenninikof, and Bendire, 1895). However, until the last twenty years, practically everything known to science about their biology stemmed from two main sources: the early writings of Pallas (1811), which derived from the observations of Steller, and probably Kraschenninikof, and the experiences of a small number of American ornithologists in Alaska; Turner (1886), Dall (Nelson, 1887), Littlejohn (related by Bendire, 1895), Willett (1914, 1915) and Heath (1915), well summarized by Bent (1919).

Some features of the Ancient Murrelet's biology were already clear from the account of Steller. According to Pallas, he mentioned that the birds visit their breeding sites only at night and that they stay out at sea during the day, feeding on small fish, molluscs and crabs. He desribed them as remarkable divers, apparently considering them better than other auks, although on what grounds is unclear. He also thought them peculiarly stupid because, as they arrived at the colony, they would land on people sitting on the ground and run underneath their clothes to seek shelter. The eggs were accurately described as being off-white or greyish with speckles or blotches, although the clutch size was given as one, occasionally two (it is normally two). If Steller's accounts were obtained first hand, then they were presumably based on experiences of the Commander Islands. It is nice to imagine Vitus Bering's famished, scurvy-ridden crew, wrecked on the Commander Islands after their historic voyage across the north Pacific, treating these tasty little winged morsels as manna from heaven, as they showered out of the night sky. However, I am inclined to doubt this scenario. Bering Island, where they wintered, was infested with arctic foxes, which proved a dire nuisance to Bering's crew (Ford, 1967). The chance that substantial numbers of Ancient Murrelets coexisted with large numbers of foxes seems low. The fact that Steller was able to give some account of the birds at the start of the voyage, when he could have had no experience of their breeding grounds, indicates that most of the information that he reported could have been obtained from others.

Most of the information supplied to Pallas by Steller was repeated almost word for word by Brandt (1869), which suggests that no further information had become available in the intervening century. Brandt added that the natives made the skins into beautiful coats called *parki*, but specifies that this information also derived from Steller. Greenland Eskimos have a similar use for Dovekie (Little Auk) skins (Freuchen and Salomonsen, 1958).

Steller described the Ancient Murrelet as nesting in cracks among rocks. While such sites are sometimes used, especially where terrestrial predators are present, a more typical site over most of their current range, is a burrow in the ground, dug by the birds themselves. The first accurate description of a burrow nest appears to derive from Dall (Baird et al., 1884), who found birds sitting on clutches of two eggs in burrows, "similar to those used by Fork-tailed Storm-Petrels (*Oceanodroma furcata*)" on the Chika Islets in Akutan Pass, near Unalaska in June 1872. He noted that birds collected while incubating included both sexes, correctly inferring, as Steller had done before, that the members of the pair share incubation duties.

Littlejohn was the first to describe the most startling feature of Ancient Murrelet biology, the precocial departure of the chicks at night. He spent several weeks in 1894 on Sanak Island, off the Alaska Peninsula, looking for Ancient Murrelet nests and eggs, and gave detailed descriptions of nest sites. Unfortunately he gave the interval between the laying of the two eggs as 2–3 days (it is normally 7–8 days), a mistake that was copied by Bent (1919) and hence found its way into most subsequent accounts, until corrected by Sealy (1976). Littlejohn rightly noted that several days frequently elapse between the completion of the clutch and the start of incubation, and mentioned that the species formed the chief prey of Peale's Peregrine Falcon in the area.

Further details on the departure of the young murrelets from the colony were given by Willett (1915) who clearly watched the entire process rather closely: "The old bird precedes the young to the water, generally keeping from 20 to 100 feet ahead of it. A continuous communication is maintained between the two, the frequent cheeps of the young being answered by the parent." This statement applies to some, but not all, departures, as related in Chapter 8. Willett also noted that the young chicks are never found close to the colony in the early morning, and hence must swim out to sea very rapidly after reaching the shore. Heath (1915), who visited Forrester Island, in southeastern Alaska, with Willett, was a little less specific in his observations, but rather more lyrical. His is the first indication that we are dealing with a truly remarkable natural spectacle:

> The journey of the young to the sea is one of the most interesting sights on the island, and by the aid of a lantern was witnessed on several occasions. The pilgrimage is made at night, within a day or two of hatching, and is evidently intiated by one or other of the parents who take up a position on the sea not far from the shore. Here, about midnight, they commence a chorus of calls resembling the chirp of an English Sparrow with the tremulo stop open, and in response the young, beautiful, black and white creatures, as active as young quails, soon pour in a living flood down the hillsides. Falling over roots, scrambling through the brush, or sprawling headlong over the rocks, they race at a surprising rate of speed, drawn by the all-compelling instinct to reach the sea.

Although others were to make observations on the Ancient Murrelet over the next 50 years, the comments of Littlejohn, Willett and Heath essentially defined the natural history of the species, as far as land-based observers were concerned. Ishizawa (1933), in Japan, published a description of the basic breeding biology of the species, which was extensively quoted by Austin

(1948), but because Ishizawa's work was in Japanese it was overlooked by most western authors. It was not until the late 1960s that a more rigorous, scientific approach took over, and this forms the material for the succeeding chapters. Whatever the statistics tell us, though, they cannot replace the magic of actually being, like Heath, at the centre of the young murrelets' amazing adventure. The true subjects of this book are not numbers, but feathers and flesh.

CONCLUSIONS

The Ancient Murrelet became known to science through the researches of Steller and Kraschenninicof in Kamchatka. Knowledge of the bird's biology reported via Pallas probably depended, to some extent, on the accounts of local people. Audubon's excellent plate, based on a somewhat aberrant specimen, formed the basis for many later descriptions. Further accounts of the species' biology accumulated through the descriptions of a small number of American naturalists, visiting Alaska. Most of the Ancient Murrelet colonies in the Queen Charlotte Islands did not become known to the outside world until the 1950s and later. The first comprehensive account of the species' breeding biology was made by Ishizawa (1933), and the first scientific account, by Sealy (1976). A good general account of the species' biology is given by Yu. Shibaev in Flint and Golovkin's (1990) *Birds of the U.S.S.R.*

Ancient, Xantus' and Marbled Murrelets

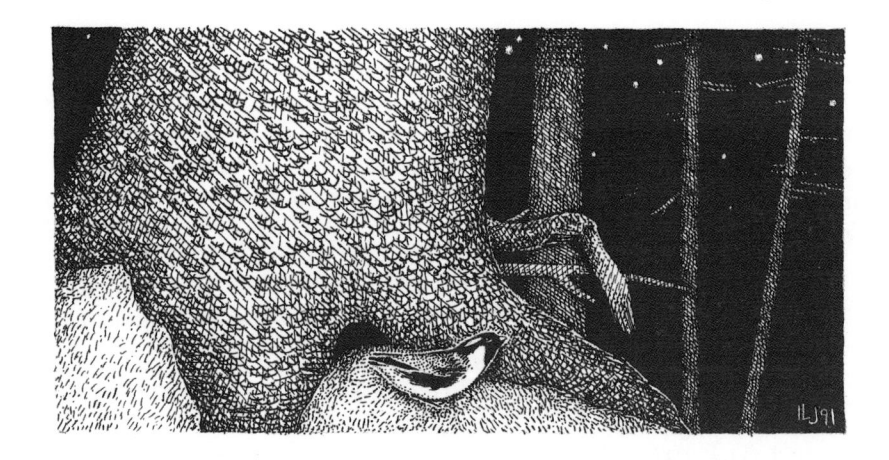

CHAPTER 3

Distribution, status and breeding habitat

Distribution and status in Asia, Alaska, the Queen Charlotte Islands, and other areas south of Alaska; their breeding habitat; the distribution and status of other Synthliboramphus *murrelets*

The Ancient Murrelet has a large range, both in terms of latitude and longitude, breeding from about 35°N in China to 61°N in Pensinskaya Gulf, on the eastern shores of Siberia, and from 131°W in the Queen Charlotte Islands, to 120°E on the coast of China's Yellow Sea. However, within these boundaries it is distributed in a thin arc, about 9000 km in length, around the northern rim of the Pacific Ocean (Figure 3.1). The information now available suggests that the species becomes progressively more abundant from China to British Columbia. However, it is very easily overlooked because of its nocturnal colony visits, and it is notoriously difficult to census, so the current numbers will certainly have to be revised as we get new information.

On the Asian side of the Pacific Ocean, it cannot be regarded as common anywhere, being only thinly distributed in small colonies separated by long distances. In the Aleutian chain it is more common, but is far outnumbered by other species of auks. Numbers there are poorly known, despite the considerable efforts of the U.S. Fish and Wildlife Service over the past 15 years. It becomes abundant once we reach the Sandman Reefs and other islands south of the Alaska Peninsula, but it is a dominant constituent of the marine bird community only in the Queen Charlotte Islands, and adjacent parts of southeastern Alaska.

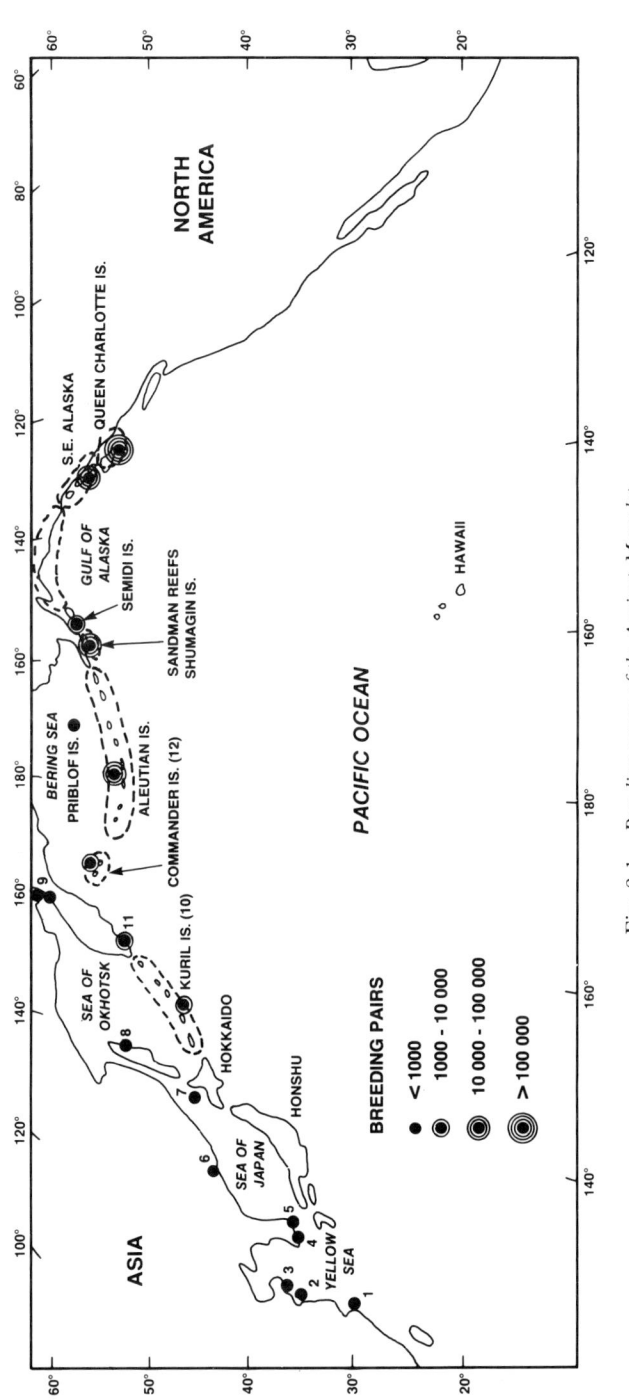

Fig. 3.1 *Breeding range of the Ancient Murrelet.*

The range of the Ancient Murrelet seems to be defined chiefly by oceanography. It occurs where mean annual surface water temperatures lie between 5° and 15°C. In the Pacific, these waters tend to be defined as subarctic (Kitano, 1981). Its distribution, defined by Udvardy (1963) as, "subboreal, pan-pacific", is similar to that exhibited by the Rhinoceros Auklet and the Tufted Puffin, although both of these species breed south to California on the east side of the Pacific, whereas the Ancient Murrelet is replaced in that area by the congeneric Xantus' Murrelet. The Ancient Murrelet's distribution is best matched by those of two unrelated genera which are similarly abundant in the Queen Charlotte Islands; the sockeye salmon and the spruce (Figure 3.2)

DISTRIBUTION AND STATUS IN ASIA

The southernmost breeding stations of the Ancient Murrelet on the west side of the Pacific are on the Yellow Sea coast of China. About 200 Ancient Murrelets breed on Chenlushan Island, in Jiangsu Province (Chen Zhaoqing, 1988). Meyer de Schauensee (1984) reports breeding on Takung Tao, off the Shandong coast, and Tso-Hsin (1987) reports it from Qingdao in Shandong and islets offshore from Shanghai Municipality. Both characterize the species as "rare".

In Korea, Ancient Murrelets seem to be somewhat more common, though still not numerous. Off Nan-do (Egg I.), near Yang-do, Neff (1956) reported, "considerable numbers on the sea near the rocks" during the breeding season. The same author mentions that local fishermen took many eggs from the island, although whether of Ancient Murrelets or other seabirds is not clear. Both Austin (1948) and Gore and Pyong-Oh (1971) mentioned that the species bred on a number of offshore islands from March to June. On Nishi Island, in northwest Korea "several hundred" Ancient Murrelets laid from mid-March to early April, while at Chibaldo Island, on the southwest coast, several hundred eggs were collected annually in 1923 (Kuroda, in Austin, 1948). The species was said to be abundant in northeast Hamgyong Pukto (close to the USSR border, in the north of Korea), where it laid in late April (Mori, in Austin, 1948), and probably nested also in Kangwon Do on the East coast. All of these observations are more than 30 years old, and the current status of the Ancient Murrelet in Korea may be rather different, but I have not been able to obtain more recent information.

Just across the border from Korea, in Peter the Great Bay, off Vladivostok, U.S.S.R., Ancient Murrelets breed on several islands. Off Verkhovskii Island, Shibaev (1987) counted 700 birds staging on 31 May and 14 June 1976. He considered that this indicated about 500 breeding pairs, and my own experience suggests that this was a reasonable guess. In the same area, the species has been recorded breeding on Russkii, Karamzin (100 pairs in the 1960s) and Klykov islands. Karamzin and Verkhovskii islands were designated natural

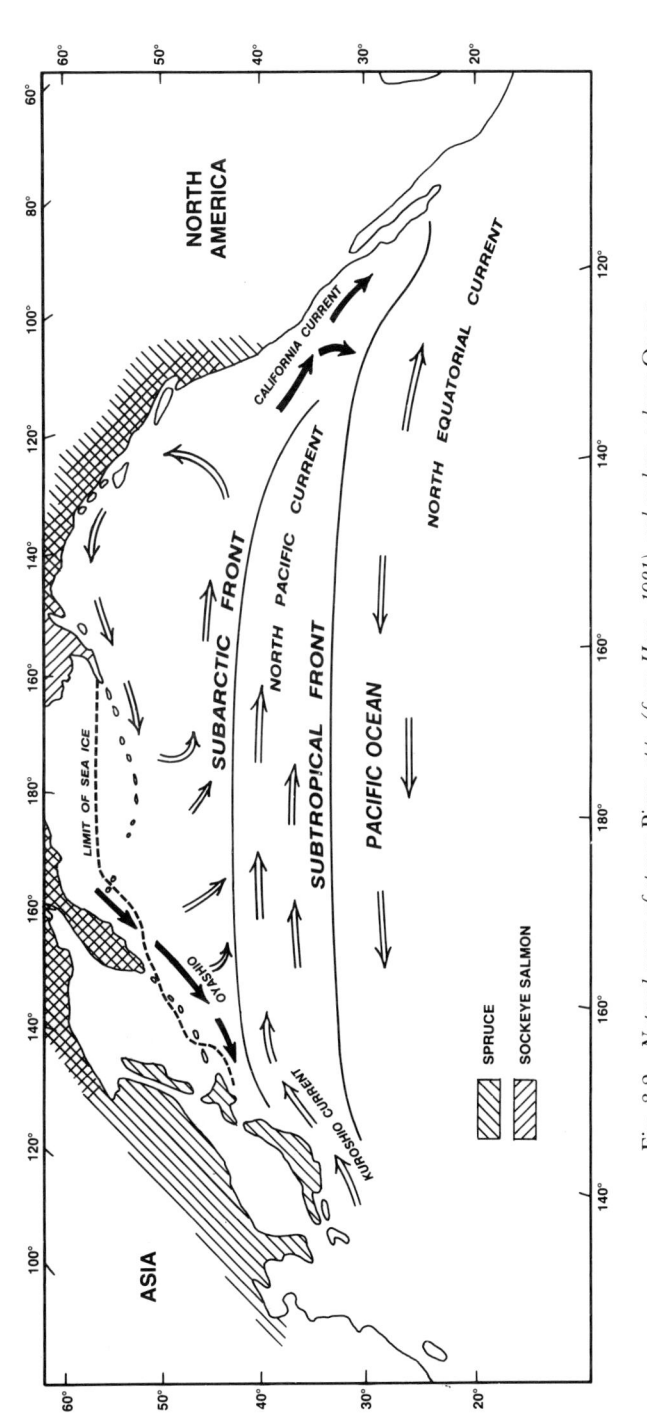

Fig. 3.2 Natural range of spruce Picea spp. (from Hora, 1981) and sockeye salmon Oncorhynchus nerka (from Childerhose and Trim, 1979), and the major North Pacific current systems.

landmarks under the authority of the Far Eastern Marine Sanctuary in 1984 (Shibaev, 1987). The latest estimate of the total population of Peter the Great Bay is 1200 pairs (Litvinenko and Shibaev, 1991). There have been several studies of the biology of the species in that area (Nazarov and Trukhin, 1985; Litvinenko and Shibaev, 1987).

Elswhere on the coasts of the Sea of Japan the status of the Ancient Murrelet is very unclear. Murata (1958) estimated 500 pairs nesting on Teuri Island, Hokkaido, Japan, "among pebbles and scanty humus", in natural cracks on a steep cliff. Later authors considered Ancient Murrelets to be much less numerous at Teuri Island (Hasegawa, 1984; Fujimaki, 1986), although there seems to be no recent information on the species' status there. It may have bred on Daikoku-Jima, off the west coast of Hokkaido (Yamashina, 1961), and there is a single breeding record from Sangan-Jima, off northeastern Honshu (Fujimaki, 1986). Ancient Murrelets occur during the summer around Tobishima, in the Sea of Japan, where they may also breed (Hasegawa, 1984). Otherwise, Litvinenko and Shibaev (1991) record simply, "scattered colonies are observed on the continental coast of the Sea of Japan from De-Kastri Bay south ...". Records from DeKastri Bay go back to the last century (Taczanowski, 1893).

Further North, in the Sea of Okhotsk, Ancient Murrelets occur during the breeding season around the Shantar Islands, although breeding has not been confirmed there (Litvinenko and Shibaev, 1991). Talan Island, not far from Magadan, on the north coast of the Sea of Okhotsk supports 5000 breeding pairs (Springer *et al.*, 1991). There are several colonies known on the coasts of Penzhinskaya Gulf, the northeastern extension of the Sea of Okhotsk, but no figures are available for population sizes (Lobkov, 1986; Vyatkin, 1986). At more than 60°N, these colonies constitute the most northerly known breeding stations for the species. Kondratiev (1991) estimated 25 000 birds in the whole of the Sea of Okhotsk, spread among 18 sites. This is the only area in which the Ancient Murrelet breeds where the sea regularly freezes over in winter.

The Kuril Islands, similar in climate and topography to the Aleutians, would seem an ideal place for Ancient Murrelets. Snow (1897), who knew the archipelago well as a result of illegal forays after fur seals (he may have been Kipling's source for his, *Rhyme of the Three Sealers*), saw the species throughout the islands in small flocks. However, recent investigations by Russian biologists have found relatively few; 3000 pairs according to Shuntov (1986), 3000 individuals according to Litvinenko and Shibaev (1991). The latter authors considered that the population of the Kurils has probably been affected by predation from introduced brown rats. Flint and Golovkin (1990) suggested that there may well be more Ancient Murrelets in the Kurils than have been located so far. Evidence of rat predation on adult Ancient Murrelets was also found on Moneron Island, one of two small colonies off the east coast of Sakhalin (Nechaev, 1986; Litvinenko and Shibaev, 1991).

In Kamchatka, Ancient Murrelets breed on the east coast on at least five islands. Starichkov Island (Island of the old men, i.e. Ancient Murrelets) is even named after them, and supports the largest population known on the Asiatic shore of the Pacific—about 6500 pairs (Vyatkin, 1986). No numbers are available for the other colonies, but they are almost certainly smaller, and some may be extinct (Flint and Golovkin, 1990). The species does not breed north of 54°N on the Bering Sea coast of Asia (Kondratiev, 1991). The population of the Commander Islands is also poorly known, although the species occurs on both Bering and Copper islands. On the former it was said to breed "only sparingly" (Stejneger, 1885), and on the latter it was "fairly frequent" (Johanson, 1961), but foxes are present on the islands, so the murrelets are probably limited to breeding in sites inaccessible to ground predators. Considering the difficulty of finding the breeding birds, unless you visit their colonies at night, the comments on population sizes probably mean little.

In summary, there are probably no more than a few hundred pairs of Ancient Murrelets breeding in China, a few thousand on the Korean Peninsula, and a similar population around the coasts of the Sea of Japan. The Sea of Okhotsk, Kamchatka and the Commander islands seem to be the species' main stronghold in Asia, but even here there are probably no more than a few tens of thousands of pairs. The wide range of the Ancient Murrelet in Asia suggests that it may have been more common formerly. Human persecution and introduced predators, especially rats, could have drastically altered its populations before any naturalist had time to assess them.

THE ALEUTIAN ISLANDS AND THE BERING SEA

My information on the status of the Ancient Murrelet in Alaska derives almost entirely from the excellent computerized database maintained by the U.S. Fish and Wildlife Service in Anchorage (Catalogue of Alaskan Seabird Colonies—Computer Archives). This database includes all of the data derived from the many surveys carried out by Fish and Wildlife Service biologists throughout the state over the past 15 years. I have also benefited from the advice of several people involved in collecting the information over the past couple of decades, and I am especially grateful to Ed Bailey, Doug Forsell, Vivian Mendenhall and Vernon Byrd for their advice.

Notwithstanding the enormous amount of survey activity that has taken place during the past two decades, the status of the Ancient Murrelet in Alaska is still very imperfectly known. While in 1989, the catalogue listed 66 known and 8 probable breeding sites, with an estimated total population of just under 110 000 breeding birds, the "best guess" for the total population of the state was 800 000 birds (V. Mendenhall, pers. comm.). Likewise, recent surveys at sea in the western Aleutians suggested that in the region of 100 000 Ancient Murrelets were present in an area thought, on the basis of colony counts, to support about 13 000 breeding birds (D. Forsell, pers. comm.). Even allowing for a substantial non-breeding population being included in the observations at sea, the comparison suggests that the colony counts had greatly underestimated the numbers present. These discrepancies illustrate how much more remains to be done to elucidate the status of the Ancient Murrelet in Alaska.

Small numbers of Ancient Murrelets occur in the Bering Sea in summer, being seen regularly around the Pribilof Islands (G. L. Hunt, pers. comm.), and occasionally around St Lawrence Island, where Sealy et al. (1971) described its status as "presumed to breed, but no definite record". The same can be said for the Pribilofs.

Ancient Murrelets occur throughout the Aleutians, but they are probably most abundant at present in the Fox Islands, at the eastern end of the chain, where at least six islands each support a thousand or more breeding pairs (Table 3.1). Elsewhere in the Aleutians, the largest colonies are on Buldir

Table 3.1 *Estimates of Ancient Murrelet populations breeding in Alaska (all numbers refer to breeding pairs).*

Area	Population Estimate	Source
BERING SEA		
Pribilof Is.	Probable	G. L. Hunt (pers. comm.)
St. Lawrence I.	Probable	Sealy et al. (1971)
ALEUTIAN ISLANDS		
Near Islands	Possibly breeds	Sowls et al. (1978)
Rat Islands		
Buldir I.	4000–5000	Byrd and Day (1986)
"Tomfredof" I.	1000	U.S. FWS* (1989), unpubl.
7 other sites	>190	Computer Archive
Andreanof Islands		
Koniuji I.	5000	Bailey and Trapp (1986)
Round I.	500	
7 other Is.	Prob. >1000, abundant on 2	U.S. FWS (1989)
Is. of the Four Mountains		
Chagulak	2500	Bailey and Trapp (1986)
Fox Islands		
Emerald I.	1500	
Egg I.	2500	
Wislow I.	1500	
Huddle Rocks	1000	
Baby Islands (5 sites)	1100	
Kaligagan I.	500	Nysewander et al. (1982)
Poa I.	500	
Aiktak I.	500	
Vesvidof I.	2500	
Ogchul I.	500	
Cape Izigan, Akutan I.	200–500	
15 other sites	>1000	U.S. FWS (1989)

Area	Population Estimate	Source
WESTERN MAINLAND OF ALASKA		
Sandman Reefs, Sanak Is. and Pavlof Is.		
Jude I.	500	Bailey and Faust (1980)
Hunt I.	5000	
21 other islands	Abundant on 8	U.S. FWS (1989)
Shumagin Is.		
Castle Rock	15 000	
Hall I.	1500	Moe and Day (1977)
Peninsula I.	1500	
5 other sites	Abundant on 4, esp.	U.S. FWS (1989)
	Karpa, Haystack	Bailey (1978)
South coast of Alaska Peninsula		
Spitz I.	Thousands	
Atkulik I.	"Thousands"	Bailey and Faust (1981)
5 other sites	Present	
Semidi Is.		
Kateekuk I.	1250	
Anowik I.	1250	Hatch and Hatch, 1983
4 other sites	>500	
Shelikof Strait/Cook Inlet		
4 sites	>300	
Kodiak I.		
East Amatuli I.	>300	U.S. FWS (1989)
1 site	Probable	
Gulf of Alaska		
4 sites	Present	
SOUTHEAST ALASKA		
St. Lazaria I.	750	Nelson et al. (1987)
Forrester I.	30 000	DeGange et al. (1977)
6 other sites	>100	U.S. FWS (1989)

* U.S. FWS, U.S. Fish and Wildlife Service.

Island, in the Rat Islands, on Koniuji Island, in the Andreanof group, and on Chagulak Island, among the Islands of the Four Mountains group (Figure 3.3). None of these estimates are likely to be better than orders of magnitude, because no detailed censuses have been carried out (E. Bailey, pers. comm.). Numbers must have been much higher in the past (Lensink, 1984), because foxes were introduced onto most of the Aleutians from 1750 onwards, and increasingly after the purchase of Alaska by the United States, to encourage fur trapping (Jones and Byrd, 1979). The seabirds acted as the food supply for the foxes, and when these had been exhausted the "farmers" sometimes introduced ground squirrels, or other rodents, which continued to affect the habitat after the foxes had died out, or been removed. The introduction of foxes to Sanak Island, where Littlejohn found Ancient Murrelets abundant in 1894, seems to have exterminated the birds. None breeds on the island today. Existing populations of Ancient Murrelets in the eastern Aleutians are all on fox-

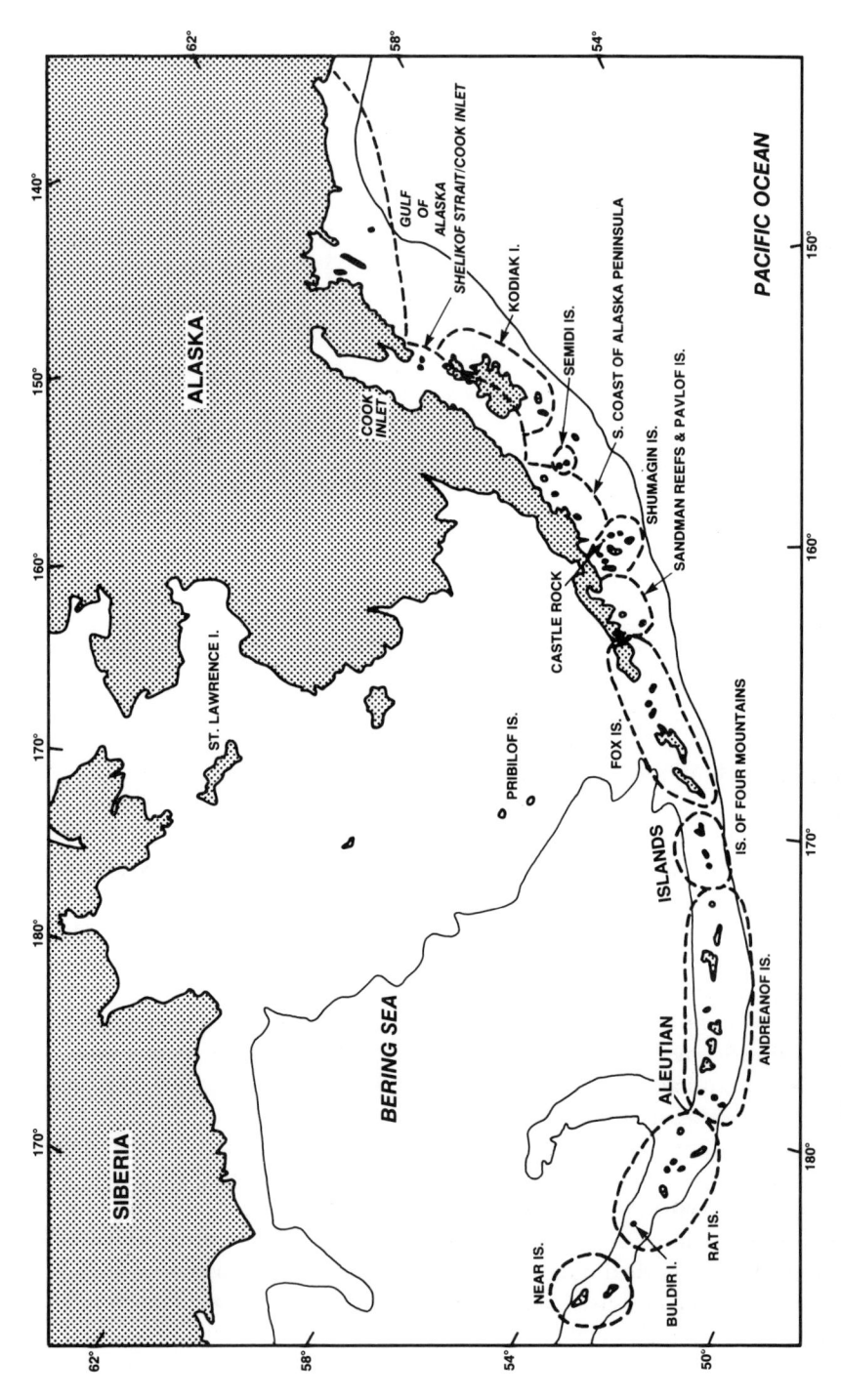

Fig. 3.3 Breeding sites of Ancient Murrelets in the Aleutian Islands and the Gulf of Alaska.

free islands (Nysewander *et al.*, 1982). The large population of 4000–5000 pairs on Buldir Island may be representative of what was found previously throughout the Aleutians (Byrd and Day, 1986).

THE ALASKAN PENINSULA AND GULF OF ALASKA

Around the Alaskan Peninsula Ancient Murrelets occur only on the south coast, where they breed on all the major island groups; the Sandman Reefs, and the Pavlof, Shumagin and Semidi islands. Few counts are available from the Sandman Reefs, but the species is common on 6 of the 14 occupied islands. The largest colony in this area where numbers have been estimated is on Castle Rock, in the Shumagins, where Moe and Day (1977) estimated 15 000 pairs. This is the largest colony reported from any treeless site, although Amagat Island may support more (E. Bailey, pers. comm.). Two other colonies in this group are believed to support over 1000 pairs, and the species is "abundant" on five others. "Thousands" occur on Spitz Island, to the east of the Shumagins, and on two islands in the Semidi group. As in the Aleutians, numbers of murrelets breeding in these islands were greatly affected by introduced foxes, which were "farmed" on most of the islands. Where the foxes have died out, or been removed, Ancient Murrelet populations will probably recover (Bailey 1978, Bailey and Faust, 1980).

To the east of the Semidi Islands, in the Cook Inlet/Kodiak Island area, Ancient Murrelets appear to be much less common than further west. Only ten sites are listed between Shelikof Strait and South-East Alaska, and none appears to be very large. There is no obvious reason for this, except that small offshore islands are somewhat less frequent in this area. However, those that are available do not support particularly large colonies. The species is uncommon in the Barren Islands, which support good populations of other seabirds, and is absent altogether from the remote Middleton Island, in the centre of the gulf (a former fox farm, and currently supporting many rabbits). On the mainland shores of Shelikof Strait an abundance of grizzly bears has been proposed as a possible reason for the lack of breeding seabirds, other than cliff-nesters (Bailey and Faust, 1984). Certainly bears could easily dig up burrows, and have been seen doing so on St. Lazaria Island (Willett, in Bailey and Faust, 1984), and the Alaska Peninsula (E. Bailey, pers. comm.). Black bears do this to storm petrels occasionally in the Queen Charlotte Islands (Drent and Guiget, 1961).

The Ancient Murrelet population in southeastern Alaska is practically all concentrated on Forrester Island, at the most southerly tip of the Alaska panhandle. The only other colony of any size is on St. Lazaria Island, where Willett (1914) found eggs. All visitors to Forrester Island have described the numbers of Ancient Murrelets as very large. The survey by DeGange *et al.* (1977) provides the best estimate of the population (30 000 pairs), but even this was, at their own admission, quite crude. It would be very interesting to know whether numbers on Forrester Island have decreased, as they have on Langara Island, on the other side of Dixon Entrance.

THE QUEEN CHARLOTTE ISLANDS

Ancient Murrelets in the Queen Charlotte Islands are concentrated in two areas: off the west coast of Graham Island, in the north, and off the east coast of Moresby Island, in the south (Figure 3.4). Both of these areas supports about 120 000 breeding pairs. The northern population is entirely concentrated in three large colonies, on Langara, Frederick and Hippa islands. The southern group is spread over at least 17 islands. An additional 10 smaller colonies, comprising about 20 000 breeding pairs, occur off the west side of Moresby Island (Table 3.2), the largest being on Helgesen Island (7700 pairs; Rodway, 1991).

The colonies off Graham Island have been known since the early 1900s, but no attempt was made to census them until 1981. Nevertheless, Spencer Sealy made a retrospective estimate of the size of the Langara Island colony for the period when he was working there in 1970 and 1971 (Vermeer *et al.*, 1984). Prior to that, only general statements, such as "astronomical" (Beebe, 1960), "immense numbers", or "thousands" (Drent and Guiget, 1961), were available. We have an even shorter record of the South Moresby colonies, which were unknown to outsiders until the 1960s, although mostly familiar to the local Haida. For most Ancient Murrelet colonies in the archipelago, even orders of magnitude were uncertain until the recent Canadian Wildlife Service (CWS) surveys. Consequently, we have only a few clues about whether populations are increasing, stable or declining.

At Langara Island, Sealy estimated 80 000–90 000 breeding pairs of Ancient Murrelets in 1971 (Vermeer *et al.*, 1984), but even by then, many formerly occupied areas were deserted (Nelson and Myres, 1976; Sealy, pers. comm.). By 1981 the estimated population had fallen to 25 700 pairs (Rodway *et al.*, 1990). Another census in 1988 provided a similar estimate, of 24 100 pairs, although the occupied area had contracted further (Bertram, 1989). Early accounts suggested that the occupied area was much more extensive than was found in the 1980s (Figure 3.5), and everything points to a dramatic reduction in the population of what was probably, at one time, the largest colony in the Queen Charlotte Islands, perhaps in the world (Nelson, 1990).

One interesting feature of the declining murrelet population on Langara Island is that it appears to be concentrating more and more. When Rodway *et al.* (1983) surveyed it in 1981 they found a relatively high density of 840 burrows/ha, averaged over the whole occupied area (101 ha). Bertram, in 1988, found an average of 1358 burrows/ha; higher than any other colony in the Queen Charlotte Islands. Estimates of average density are obviously

Distribution, status and breeding habitat   31

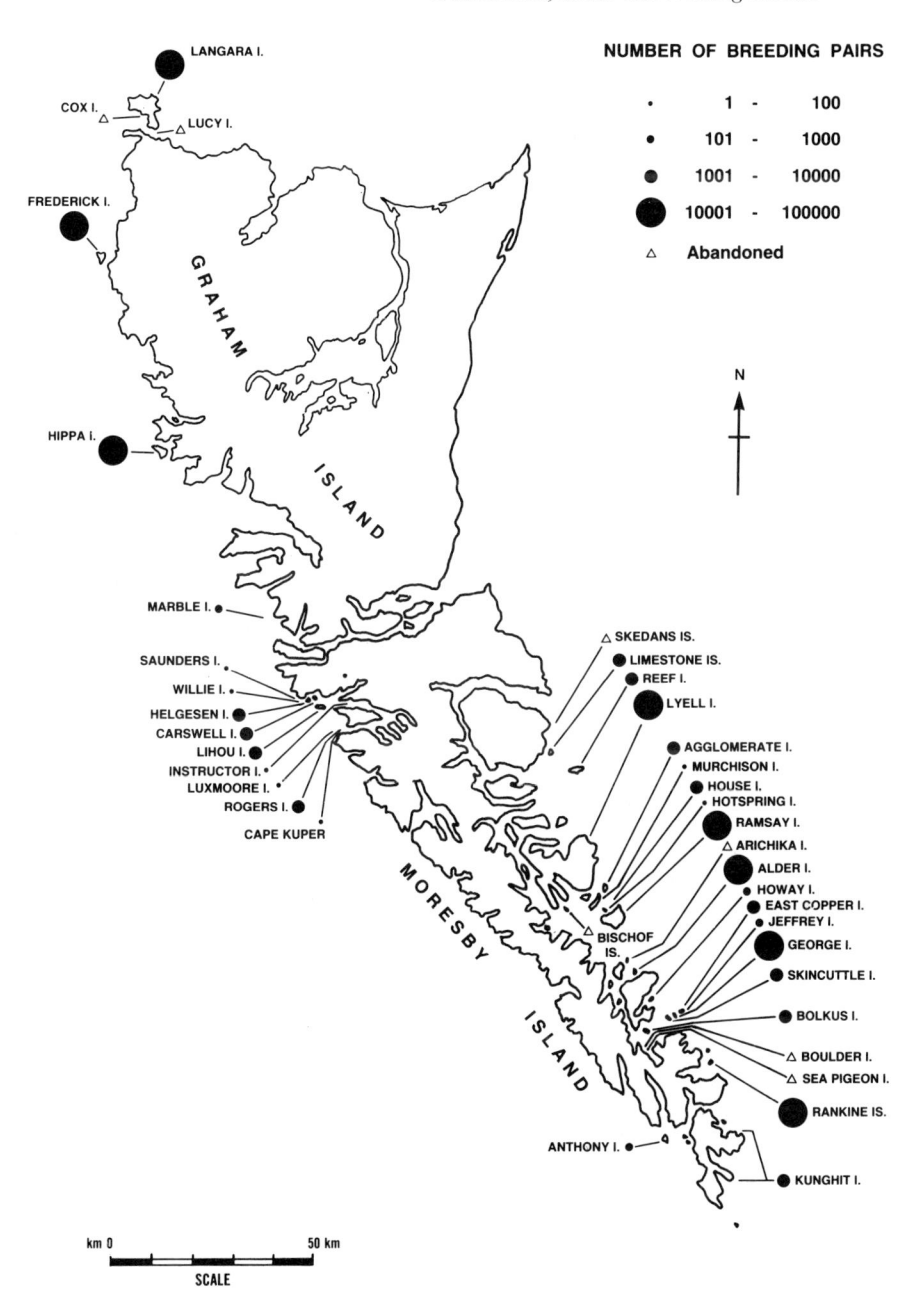

Fig. 3.4 *Breeding sites of the Ancient Murrelet in the Queen Charlotte Islands.*

Table 3.2 *Breeding populations of Ancient Murrelets in the Queen Charlotte Islands (from Rodway, 1991).*

Locality	Population (pairs)	Year of survey
WEST COAST OF GRAHAM ISLAND		
Langara I.	24 000	1988
Frederick I.	68 000	1980
Hippa I.	40 000	1983
Marble I.	1000	1977
WEST COAST OF MORESBY ISLAND		
Saunders I.	50	1986
Helgesen I.	7700	1986
Willie I.	10	1986
Carswell I.	1700	1986
Instructor I.	760	1986
Lihou I.	6500	1986
Luxmoore I.	1000	1986
Rogers I.	1700	1986
Cape Kuper, Moresby I.	10	1986
Anthony I.	200	1985
EAST COAST OF MORESBY ISLAND		
Kunghit I.	8800	1986
Rankine I	26 000	1984
Bolkus I.	9900	1985
Skincuttle I.	2200	1985
George I.	11 600	1985
Jeffrey I.	1000	1985
East Copper I.	4400	1985
Howay I.	300	1985
Alder I.	14 400	1985
Ramsay I.	18 200	1984
Hotspring I.	6	1984
House I.	2600	1984
Murchison I.	20	1984
Agglomerate I.	2200	1985
Dodge Point, Lyell I.	10 700	1982
Reef I.	5000	1985
Limestone Is.	1500	1983

affected by the choice of colony boundary. If some unoccupied areas are included the density will be lower, but compensated for in the population estimate by the increased area of the colony. In this case a difference in the interpretation of where the colony boundary was cannot explain the differences in density, because the mean density found by Bertram was greater than the maximum density found by Rodway *et al*. We have to conclude that, despite the contraction taking place in the colony area, and despite a very low rate of burrow occupancy (26% in 1981), the Ancient Murrelets continued to dig burrows.

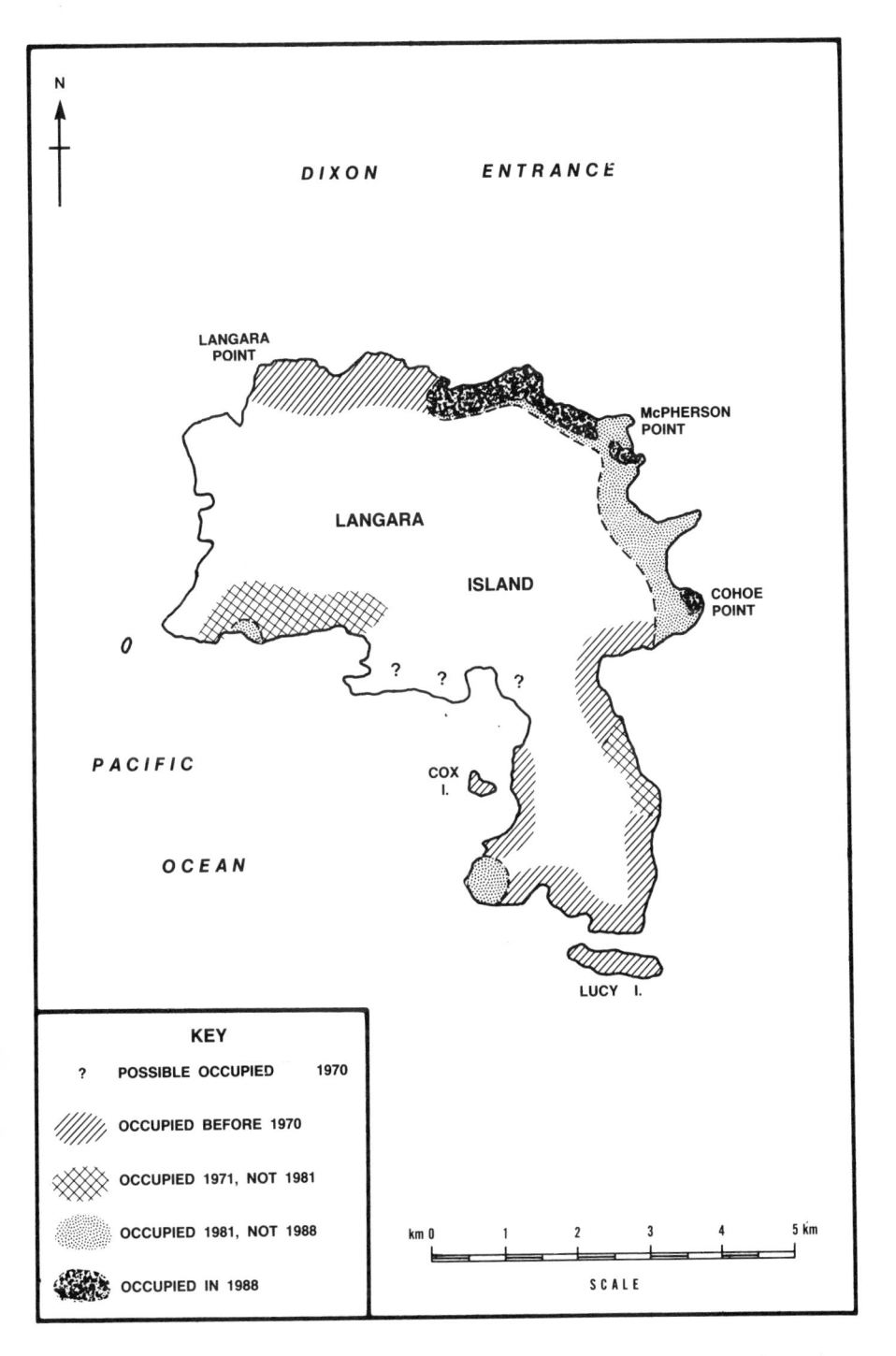

Fig. 3.5 *Changes in the distribution of breeding Ancient Murrelets on Langara Island since the 1950s.*

The size of the original Ancient Murrelet population of Langara Island is impossible to know, but we can make an educated guess. The extent of coastline suitable for Ancient Murrelets is about four times that occupied in 1981. If we assume that the density of burrows in 1981 was representative of earlier times, and that the proportion of burrows occupied would have been similar to the average for Queen Charlotte Island colonies (63%), we obtain an estimate of more than 200 000 pairs. This is more than three times the size of the largest Ancient Murrelet colony at present (Frederick Island), and certainly sufficient to have prompted the kind of hyperbole used to describe it by Beebe, Drent and Guiget, and others.

The decline in the number of murrelets on Langara Island has been accompanied by a dramatic decline in the numbers of Peregrine Falcons breeding there, which fell from about 20 pairs in the 1950s to 5 or 6 pairs during 1968–73, and have since remained about the same (Nelson and Myres, 1976; Nelson, 1988, 1990). The Ancient Murrelet is the main prey of the peregrine at Langara Island (Beebe, 1960; Nelson, 1977), and it is very tempting to link the two population declines. However, the timing does not fit very well.

The apparently steep decline in the number of murrelets between 1970 and 1981 was not reflected in a similar decline in peregrines. The peregrine decline seems to have been simultaneous with the worldwide peregrine crash, which took place in the late 1950s and 1960s, following the introduction of chlorinated hydrocarbon pesticides (Newton, 1979; Ratcliffe, 1980). Levels of chlorinated hydrocarbons in peregrine eggs from Langara Island taken between 1967 and 1972 (geometric mean 7.8 ppm DDE, with 5/10 eggs at >15 ppm; Peakall *et al.*, 1990) were close to those proven to cause reduced reproductive success in other peregrine populations (Nelson and Myres, 1976; Newton, 1979; Peakall *et al.*, 1990). Eggshell thickness was also reduced by 13% compared to pre-1947 levels (Anderson and Hickey, 1972). Nelson and Myres (1976) obtained evidence that reproductive success during 1968–73 was lower than when Beebe had studied the Langara peregrines, in the 1950s. All these observations suggest that the reduction in Langara peregrines may have been, at least partly, the result of chlorinated hydrocarbon contamination. Levels of DDE in four peregrine eggs taken from Langara Island in 1986 showed a marked reduction in contamination (maximum 6.6 ppm; Peakall *et al.*, 1990), compared with the situation in the late 1960s. Despite this improvement, there is no sign of any increase in the peregrine population. Consequently, even if the decline in the Ancient Murrelet population did not precipitate the peregrine decline, it is very possible that it has been responsible for the lack of a subsequent recovery. Nelson (1990) has suggested that even the present level of the peregrine population on Langara Island can only be maintained because the falcons are preying partly on Ancient Murrelets feeding in Dixon Entrance but originating from the large colonies on Forrester and Frederick islands.

Nelson and Myres (1976) suggested that the Ancient Murrelet population declined on Langara Island because there was a reduction in the availability of food, either because of pesticide pollution, or because of changes in the marine current systems controlling local productivity. Levels of organochloride pesticide residues (DDE 0.4–2.0, dieldrin 0.004–0.014 ppm wet weight) in a small sample of Ancient Murrelet eggs taken at Langara Island in 1968 were actually lower than those observed in 1970 and 1971 in the eggs of a variety of other seabirds, including storm petrels, which have diets similar to that of the Ancient Murrelet (Elliot et al., 1989). This makes it unlikely that pesticides were involved. The levels found were similar to those seen in a small sample of eggs from Buldir Island, in the western Aleutians (Ohlendorf et al., 1982). More recent analyses of deserted eggs taken from Reef Island in 1986 showed that levels of DDE had fallen only slightly, to 0.3–1.5 ppm, and dieldrin to 0.003–0.006 ppm. Levels of the compound beta-HCH were higher in Ancient Murrelets than in any other species studied in the 1980's (Elliot et al., 1989). Beta-HCH is a constituent of pesticides used principally in Asia. Its presence in Ancient Murrelets in the Queen Charlotte Islands suggests that it is being transported across the Pacific by ocean currents. Thus developments half a world away may be having repercussions even in the wilderness of South Moresby!

The possibility that oceanographic changes are involved in the decline of Ancient Murrelet populations is hard to discount, because there is very little information. However, long-term studies in California (Ainley and Boekelheide, 1990) and Alaska (Hatch et al., in press) do not suggest directional changes in oceanographic events affecting those areas, which bracket the Queen Charlotte Islands.

Studies by Bertram (1989), showed that ship rats, introduced on Langara Island some time before the end of the Second World War, had killed many adult Ancient Murrelets in their burrows. Evidence of rat predation on Langara had been noted earlier by Campbell (1968), Sealy (1976) and Rodway et al. (1983). Bertram found murrelet bones in 29% of all burrows searched, and they were most common in parts of the colony that had been abandoned. Bones are rarely found in burrows in colonies where rats are absent (Rodway et al., 1988; Bertram, 1989; pers. obs.). There is a strong suggestion that rats were responsible for the decline in Ancient Murrelets at Langara Island, at least in recent years, although we do not know when the rats arrived on the island. It is worth noting that at Murchison Island, where rats are also present, there are only a handful of breeding Ancient Murrelets, although the nearby and very similar House Island supports more than 2000 pairs. There is a possibility that the population on Murchison has been greatly reduced by the rats. Lucy Island, adjacent to Langara Island, which was described as "perforated with burrows" in 1946 (Drent and Guiget, 1961) is now abandoned, perhaps because martens were introduced (Rodway, 1991).

Although our information about population changes on Langara Island is scanty, it is still better than anything available elsewhere in the Queen

Charlotte Islands. Several small colonies in the southeast Moresby area disappeared during the last two decades. Murrelets were present on Low Island and the Skedans Islands in 1970, but were gone by 1983 (Summers, 1974; Rodway *et al.*, 1988). We heard birds calling on Low Island occasionally, while catching storm petrels there in 1985–87, so it is possible that a few still breed. It is unlikely that either of these sites would have supported large populations. It may be significant that the Peregrine Falcon sites on these two islands, active in earlier decades, were never occupied during the 1980s (K. Moore, pers. comm.).

On Boulder and Sea-Pigeon islands, in the inner part of Skincuttle Inlet, Bristol Foster found eggshells and the remains of dead adults in 1960 (Drent and Guiget, 1961), but Summers (1974) found no sign of breeding in 1971. There is a possibility that the remains found by Foster could have been brought to the islands by predators, or scavengers. If Ancient Murrelets did nest there, introduced raccoons, sometimes present on Boulder and Sea-Pigeon islands, may have been responsible for their abandonment by Ancient Murrelets.

Summers (1974), on his survey in 1971, found abundant burrowing on Arichika Island and the Bischof Islands, estimating 500 pairs on each. By 1985 Mike Rodway and Moira Lemon could find no trace of these colonies during their surveys (Rodway *et al.*, 1988). The abruptness of their disappearance may be misleading; many burrows could already have been deserted when Summers visited the islands. Although these colonies have certainly disappeared recently, the cause of their desertion is unknown.

The only other colonies for which population changes have been demonstrated are the East and West Limestone Islands, in the South Moresby area. Summers (1974), found burrows covering both islands, except for a small area on the east island. He estimated the total population at over 5000 pairs. In 1983 the CWS survey team estimated about 1500 pairs on the two islands, and in 1989 we obtained a similar estimate, with only a handful remaining on the west island. The cause of the decline is unknown, but the remains of predator kills, always common on Ancient Murrelet colonies, are especially abundant on East Limestone Island. In some cases we found burrows that had been excavated and the adult birds eaten. Raccoons visit the island from time to time and could have been responsible for some of the mortality in the past. In the 1990 and 1991 breeding seasons, raccoons were definitely active in the colony, digging up burrows, killing adults and eating eggs (Laskeek Bay Conservation Society, pers. comm.).

Our very fragmentary knowledge of changes in Ancient Murrelet populations suggests that introduced predators, whether arctic foxes in Alaska, rats in the Kuril Islands and at Langara Island, or raccoons elsewhere in the Queen Charlotte Islands, have been the dominant factor in determining population trends. This is true for many seabirds breeding on remote islands (King, 1984; Atkinson, 1985). While things seem to be improving in the Aleutians, where foxes have been eradicated from some islands to protect the distinctive Aleutian race of the Canada Goose, conditions may continue to deteriorate in the species' present stronghold, the Queen Charlotte Islands. If we want to maintain current populations there we shall need to watch out for any development that might spread rats to areas that are currently rat-free, especially the large colonies on Hippa and Frederick islands.

OTHER AREAS SOUTH OF ALASKA

Outside the Queen Charlotte Islands, there are only two definite records of Ancient Murrelets breeding in North America south of Alaska. A nest with eggs was found by Hoffman (1924) on Carroll Island, Washington State, and another was reported from the Moore Group, off the mainland coast of Hecate Strait in 1970 (Campbell *et al.*, 1990). There have been no subsequent records from Washington, but small numbers are seen offshore during summer and a fledgling was recorded in 1978. A very small breeding population may exist there still (Speich and Wahl, 1989). Small numbers are also seen off the northern part of the West coast of Vancouver Island in summer, and breeding is possible there but no evidence for it exists. Family parties, including small chicks, are sometimes seen in the southern part of Queen Charlotte Sound, but these may well have originated from the Queen Charlotte Islands, because dispersal away from the breeding sites is very rapid (Chapter 8).

There is no evidence that Ancient Murrelets breed south of the Queen Charlotte Islands at present. However, the fossilized remains of more than one individual of the species have been found in Pleistocene deposits in California, where they are now only rare stragglers (Howard, 1949). This does not imply breeding, but it does suggest that Ancient Murrelets may have been more numerous in the southern parts of their range during the Pleistocene. At that period, they may well have nested further south, especially as some of their breeding sites in the Queen Charlotte Islands would have been denied to them by permanent ice cover (Clague, 1989).

BREEDING HABITAT

Both Pallas (1811) and Brandt (1837) described the breeding sites of the Ancient Murrelet as "cracks in the rocks on rocky shores". Both give Steller as authority, and he may have been referring to the situation on Bering Island, where there are no trees, and where arctic foxes are native. In the Queen Charlotte Islands, in South-East Alaska, in some places in Peter the Great Bay, and probably also in the Kuril Islands, Ancient Murrelets nest under forest. However, from Kamchatka through the Commander and Aleutian islands, and as far east as the Gulf of Alaska, most of the islands on which they breed are treeless. Where forest is available, it appears to be the species' preferred breeding habitat (Vermeer et al., 1984). Where it is absent, Ancient Murrelets usually pick the most densely vegetated area available, provided that it is not waterlogged. On Buldir Island, they occupy the lowland tall-plant complex, which grows about 1 m high (Byrd and Day, 1986). In the eastern Aleutians they are found in *Elymus/Calamagrostis* grassland and mixed *Elymus* and umbelliferae. Their burrows are also found in the foundations of abandoned native houses (Nysewander et al., 1982). Tussock grasslands is probably their main breeding habitat throughout much of their Alaskan range (Bendire, 1895; Nysewander et al., 1982), but they also breed on some small islands practically devoid of vegetation, and there they must make do with cracks in the rocks; they are common in such habitat on islands off the Alaska Peninsula (E. Bailey, pers. comm.).

Steller's account of breeding in rock crevices appears to be fairly typical of nesting sites found over most of the species' range in Asia. In China, Chen Zhao-Qing (1988) describes Ancient Murrelets as nesting "in natural rocky holes up to 60 cm long". In Korea, Ishizawa (1933) says simply "makes a slight hollow in the ground, lined with a small quantity of dead grass and fallen leaves", while in Japan, Murata (1958) describes nesting in natural rock crevices 30–50 cm deep. In Peter the Great Bay nests are placed in gaps between rocks, under boulders on the beach, or in burrows dug in soil (Litvinenko and Shibaev, 1987), while a few are in crevices in cliffs so that the chicks jump directly to the water (Shibaev, 1978). On Starichkowa Island most sites are in burrows on steep grassy slopes (Vyatkin, via N. Litvinenko, pers. comm.).

Burrows dug by the birds themselves seem to be the most typical nest sites where there is a sufficient depth of soil. In forest habitat, burrows are tunnelled under the base of trees, stumps, or fallen logs, and may penetrate fissures in the underlying rocks (Drent and Guiget, 1961). Outside forest they make use of rock crevices, or burrows made among the roots of grass tussocks. Bendire (1895) described such burrows on Sanak Island, in the Aleutians, in detail:

> Under these almost solid bunches [of grass] the murrelets would force their way, leaving only a slight hole in the mass, which was usually very hard to detect . . . a shallow cavity about 5" in diameter and 2–3" deep was scratched out, and this was nicely lined with blades of dry grass.

Ancient Murrelets no longer occur on Sanak Island, thanks to the ubiquitous introduced foxes.

Bendire mentioned that in some instances no nest cup was made, the eggs being laid on rock, sand, or muddy soil, sometimes even on ice. Whatever site is chosen, Ancient Murrelets prefer to incubate in the pitch dark. Consequently, the access tunnel is usually crooked, the nest chamber being placed out of sight from the entrance.

THE OTHER *SYNTHLIBORAMPHUS* MURRELETS

The Japanese Murrelet breeds only in Japan, on small offshore islands. It is the rarest of the world's auks, and is the only one listed as threatened on a world scale in the *World Checklist of Threatened Birds* (Collar and Andrew, 1988). It prefers warmer waters than does the Ancient Murrelet, breeding in the subtropical and warm-temperate oceanographic zones of the Kuroshio and Tsushima currents, to the south of the main island of Honshu (Hasegawa, 1984).

My information on the distribution of the Japanese Murrelet comes mainly from Higuchi's (1979) excellent review. At one time it bred close to Tokyo, in Kanagawa Prefecture, but has almost certainly disappeared from that area. It has probably gone also from Mikomoto Island, off the Izu Peninsula, and from Yamaguchi Prefecture, at the western extremity of Honshu (Figure 3.6). The main stronghold of the species is the Izu Islands chain, which stretches south from Yokohama Bay to about 32°N. In this group it has been thought to breed on at least 11 islands, with breeding definite on four. Many of these records are now fairly old. The only site where breeding has been proven since 1970 is Sambondake Island, which was used as a bombing range by the U.S. Air Force in the 1950s (Austin and Kuroda, 1953). Apparently the murrelets survived the onslaught.

Elsewhere, Japanese Murrelets breed on Mimiana Island, off southern Honshu, and on Okinoshima and Hashira islands, off the north coast of Kyushu, on Biro Island, off the east coast of Kyushu, and on Hanaguri in the Danjo Islands. About 400 birds are thought to breed on Okinoshima (Higuchi, 1979), and 150–200 on Mimiana, off Central Honshu (Hasegawa, 1984), but no information is available on the size of other populations. Hasegawa (1984) estimated a total population of 1650 birds. There are several records of Japanese Murrelets off the southern parts of the Pacific coast of the U.S.S.R., including southern Primoriye, southern Sakhalin and the southern parts of the Kuril chain (Flint *et al.*, 1984; Kostenko *et al.*, 1989). However, breeding has not been substantiated and the observations may have involved non-breeders, or post-breeding dispersers. A partially feathered juvenile found in Peter the Great Bay in 1984 (Nazarov and Shibaev, 1987) need not have been reared locally, because some family parties of murrelets travel long distances after leaving the breeding colony (Chapter 8).

On Koyashima Islet, off the northern coast of Fukuoka Prefecture, Takeishi (1987) found the remains of 77 adult Ancient Murrelets that had apparently been killed by Brown Rats. He estimated that 270 had been killed altogether. It appeared that the murrelets had been completely extirpated from the island. In the Izu Islands, and probably elsewhere, the birds are harassed by fishermen. Like the Ancient Murrelet, they would be very vulnerable to introduced rats, or other mammalian predators. Dr Hito Higuchi, of the Japan Wild Bird Society told me in 1990 that he believed there could be less than a thousand

40 *A global glimpse of the Ancient Murrelet*

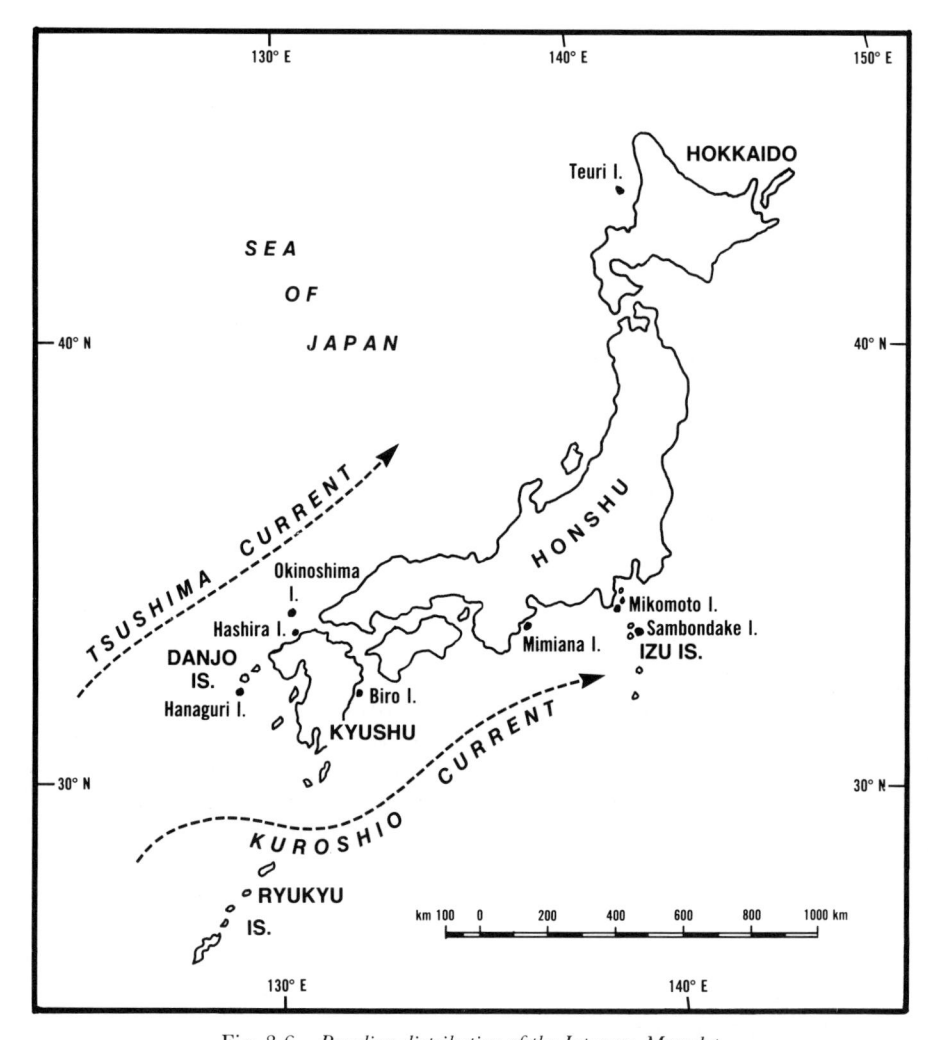

Fig. 3.6 *Breeding distribution of the Japanese Murrelet.*

pairs left. This makes the Japanese Murrelet by far the most endangered species of auk, and among the most endangered seabirds in the Northern Hemisphere. Proper protection for their breeding sites is urgently needed to ensure that the species survives.

Nesting sites of the Japanese Murrelet are very similar to those of the Ancient Murrelet. According to Austin and Kuroda (1953), they are found in rock crevices or under boulders. Higuchi (1979) describes a wide variety of additional sites, including burrows 50–100 cm long under the roots of holly trees, hollows at the base of grass tussocks, and old petrel burrows. In fact, the description of sites occupied by the Japanese Murrelet embraces practically all the types of site used by the Ancient Murrelet over its hugely greater geographical range.

Populations of Craveri's and Xantus' Murrelets are also small and localized. Both are included as 'near threatened' by Collar and Andrew (1988). Craveri's Murrelet breeds only in Baja California, mainly on islands in the Sea of Cortez. Its status is very poorly known, but flocks of a hundred or more birds are not uncommon during the breeding season (D. Croll, pers. comm.). The total population of Craveri's Murrelet has been estimated as 5000 pairs (Springer *et al.*, in press).

Craveri's and Xantus' Murrelet are very alike in appearance, and were thought to be conspecific until both were found together on the San Benito Islands, off the Pacific coast of Baja California (Jehl and Bond, 1975); however, this still requires confirmation, according to DeWeese and Anderson (1976). Xantus' Murrelet breeds on the coast of California, as far north as Point Conception. It avoids the cool upwelling areas, which centre on Monterey and San Francisco, while breeding. The main centres of the population are Santa Barbara Island, off southern California (several thousand pairs), and the outer coast of Baja California (10 000–20 000 pairs; Springer *et al.*, in press). The population of Santa Barbara Island was much reduced by cats early in the 20th century, but has since recovered, after cats were removed (Hunt *et al.*, 1980).

Unlike the Ancient and Japanese Murrelets, the other two species of *Synthliboramphus* do not appear to dig burrows for themselves, perhaps because the islands on which they breed are very rocky. The Xantus' Murrelet finds natural cavities among rocks, or under dense vegetation (Bent, 1919), usually quite shallow, so that the incubating bird can be seen from the entrance (Murray *et al.*, 1980). The breeding sites of Craveri's Murrelet are similar; rock crevices, "at arms length" (Bent, 1919).

SUMMARY

The genus *Synthliboramphus* comprises four species, but three of these have a very local distribution, being adapted to warmer waters at the fringe of the Ancient Murrelet's mainly subarctic range. Although their range was one of the last areas of the globe to be affected by industrialization, all Ancient Murrelets have almost certainly been much affected by human activities. In particular, they are very vulnerable to ground predators introduced onto their breeding islands. Their nocturnal habits protect them against most avian predators (although the Peregrine Falcon remains a major predator throughout their range), but their tendency to sit about on the ground, and their somewhat clumsy behaviour on land, make them easy targets for mammalian predators.

Ancient Murrelets occur in large colonies in the Queen Charlotte Islands and adjacent South East Alaska, but over the rest of their range their colonies are only moderate in size, by the standards of most seabirds. Their distribution on the Asian side of the Pacific, with scattered groups of 100–1000 pairs, is reminiscent of the way in which the other three *Synthliboramphus* species are distributed. Their ability to survive in small, isolated pockets has probably been important in allowing them to maintain themselves across a sprawling geographic range.

CHAPTER 4

The Ancient Murrelet's year

The annual cycle of the Ancient Murrelet; variation in the timing of breeding; the effects of latitude and sea surface temperature on regional differences in breeding seasons

To provide a time frame against which the details of Ancient Murrelet biology can be viewed, I shall give a brief description of the annual cycle of the species. In the Queen Charlotte Islands, Ancient Murrelets begin to visit their colonies in March. At Lyell Island in February 1980 no birds visited the nesting slopes (J-P Savard, pers. comm.). In 1991 birds were seen on East Limestone Island in large numbers periodically from 16 March onwards (M. vanden Brink, pers. comm.). We made our earliest observations on Reef Island on 23 March 1989 and we heard birds calling in the forest that night, so the first visit by the murrelets was probably earlier. In 1971, Sealy (1976) arrived on Langara Island on 17 March, saw the first Ancient Murrelets offshore on 30 March, and first recorded birds on land on 5 April.

The date of arrival of Ancient Murrelets at their breeding grounds varies over their range. At Amchitka, in the Aleutians, the first arrivals occurred on 23 April (Kenyon, 1961). At Adak Island, also in the Aleutians, and on the Commander Islands they arrive in May (Stejneger, 1885; Dementiev and Gladkov, 1969; G. V. Byrd, pers. comm.). During the pre-laying period, the birds are present on land only at night. They gather in large numbers just offshore from their colonies in the late afternoon, while they wait to fly in. This behaviour is sometimes termed "staging", and the fairly discrete area where they wait is known as the "staging area". However, the term "staging area" is also used for feeding grounds used by migrant birds en route from summer to winter quarters. It also seems to imply that the birds are merely waiting to visit the colony, rather than actively making use of the area. For both these reasons I have substituted the terms "gathering" and "gathering ground" for this phenomenon.

Egg laying begins at Reef Island in the last few days of March, or the first ten days of April (Chapter 13). More than 90% of clutches are begun over a period of about 30 days, although laying has been recorded as late as the end of May. Elsewhere in the southern part of the Queen Charlotte Islands the timing of laying appears very similar to Reef Island, but at the large colonies off the northwest coast (Hippa, Frederick and Langara islands), and at the nearby Forrester Island, in South-east Alaska, laying normally begins and peaks about 5–10 days later than on Reef Island (Table 4.1).

Timing of breeding varies considerably across the geographical range of the Ancient Murrelet. Although rigorous observations of laying and hatching dates are available for only a few sites, there are frequent references in the literature to the peak periods of chick departures at different sites, and some to the peak of egg laying. At colonies where good data are available on the timing of departures, large numbers are usually observed leaving over no more than 10–15 days (Chapter 13), and hence references to large numbers of chicks departing probably identify the peak of departures to within about 2 weeks. I have summarized the information relating to the timing of departures in Table 4.1 and have used it to estimate the likely date of peak departures. I have then extrapolated backwards to the approximate date of clutch completion by subtracting 34 days from my estimates of median departure. The resulting numbers are certainly rather crude, but they tend to be fairly consistent for particular areas and hence probably provide an adequate description of regional variations in the timing of breeding.

The earliest laying occurs at the southern edge of the Ancient Murrelet's range, on the Asian side of the Pacific. In China the species has been described as nesting from March to May, while I estimate median clutch completion in Korea as 5–25 April. The Queen Charlotte Islands, South-east Alaska, and Peter the Great Bay, near Vladivostok, come next, with median dates of clutch completion ranging from 21 April to 10 May. Laying in the Aleutians, in Kamchatka and in the Gulf of Alaska is later, with median dates of clutch completion from 1 to 27 June. Surprisingly, at Shikotan Island, in the southern Kurils, and only a little north of Peter the Great Bay, laying does not occur until June, and the timing is similar to colonies in the Aleutian islands.

Latitude seems to have relatively little role in determining timing of breeding, because the northern colonies of the Queen Charlotte Islands are at the same latitude as those of the western Aleutians, which lay about a month later (Figure 4.1). A much closer relationship can be found with April mean sea surface temperatures in the breeding area (Figure 4.2). The trend can be extended by adding in the southern species of *Synthliboramphus*, which all lay earlier, allowing us to make the very approximate generalization that the peak

Table 4.1. *Evidence for timing of breeding of the Ancient Murrelet and other* Synthliboramphus *species. Numbers in parentheses in the first column refer to Figure 4.1.*

Locality	Evidence for timing of breeding	Median clutch completion	Reference
Yellow Sea (1)	Nesting March to May	(15 March)	Chen Zhao-qing (1988)
Korea (2)	Lays mid-March to mid-April	5 April	Ishizawa (1933)
Korea	Nesting March–June	?	Gore and Pyong-Oh (1971)
NW Korea	Lays mid-March to early April	1 April	Austin (1948)
NE Korea	Lays late April	25 April	Austin (1948)
Peter the Great Bay (3)	Lays in mid-April, hatching from 20 May to mid-June (peak 28–31 May)	26 April	Litvinenko and Shibaev (1987)
	Peak departures, first half of June	6 May	Shibaev (1987)
Shikotan I., S. Kurils (4)	Laying, first 10 days of June, hatching in the second half of July	15 June	Litvinenko and Shibaev (1987)
Kamchatka (5)	Peak hatch 1–2 August	27 June	Vyatkin (1986)
Talan I., Sea of Okhotsk	Laying began 23 June, peak from 28 June	1 July	Flint and Golovkin (1990)
Commander I.	Eggs, June and July	(10 June)	Dementiev and Gladkov (1969)
ALEUTIANS			
Buldir (6)	Peak departures 2nd week July	6 June	Byrd and Day (1986)
Sagchudak (7)	Many departing, 14 July	10 June	Bailey and Trapp (1986)
Chika	Birds in burrows, 2 June	?	Baird *et al.* (1884)
Sanak	Many with eggs, 23 June	?	Bendire (1895)
GULF OF ALASKA			
Spitz I. (8)	Hundreds of chicks departing, 4 July	1 June	Bailey and Faust (1984)
Semidi I. (9)	Chicks departing 15–26 July	15 June	Leschner and Burrell (1987)
Shumagin Is.	Many 2-egg clutches, 17 June, most chicks had left by 24 July	(10 June)	Moe and Day (1977)
SE ALASKA			
St Lazaria I.	Many chicks had departed by 17 June	(10 May)	Nelson *et al* (1987)
Forrester I. (10)	Peak departure of chicks, 1–10 June	1 May	Willett (1915)
	Laying early May, peak departures second week of June	5 May	Heath (1915)
	Peak departures, first week of June	30 April	Degange *et al.* (1977)
QUEEN CHARLOTTE ISLANDS			
Langara I.	Still some eggs, 16–17 June	?	Young (1927)
	Nesting mostly over by the end of June	?	Cumming (1931)

Table 4.1 cont

Locality	Evidence for timing of breeding	Median clutch completion	Reference
	Begin laying, late April, start leaving, early June	(10 May)	Beebe (1960)
	Main departures of chicks, first week of June 1947	1 May	Drent and Guiget (1961)
	Mainly single eggs on 30 April 1966	(5 May)	Campbell (1968)
(11)	Median date of first egg, 2 May 1971	9 May	Sealy (1976)*
	Median start of incubation, 1 May 1988	30 April	Bertram (1989)
Frederick I. (12)	Departures 27 May to 7 June	29 April	Carson (1981)
	Median date of hatching, 3 June	2 May	Vermeer et al. (1985)
Reef I. (13)	Median clutch completion 21–28 April	24 April	Chapter 12
Limestone I.	Many burrows contained chicks, 28–29 May	24 April	Summers (1974)
(14)	Peak departures in late May	25 April	pers. obs.
Lyell I. (15)	Median hatching 21 May 1982 ($N = 17$)	19 April	Rodway et al. (1988)
Ramsay I.(16)	Median date of laying of first egg 27 April 1984 ($N = 34$)	4 May†	Data from CWS survey
S. wumizusume			
Izu Is., Japan (17)	Breeds in February to May	(15 March)	Austin and Kuroda (1953)
Honshu, Japan	Breeds mid-February to early April	(10 March)	Higuchi (1979)
S. hypoleucus			
California	Begin laying in late January	?	Brewster (1902)
	Eggs mid-March to mid-July, first brood in March	(20 March)	Bent (1919)
(18)	Median egglaying 20 March to 18 May during 1975–1978	5 April (ex. 1978)	Murray et al. (1983)
S. craveri			
Baja California (19)	Eggs present from 6 February to 12 April	(20 Feb.)	Bent (1919)
Gulf of California	Chicks at sea from March onwards		DeWeese and Anderson (1976)

* Sealy (1976) reported first departures of chicks at Langara Island on 26 or 27 May in four years: 1947, 1952, 1970 and 1971.
† 1984 was the latest year out of six at Reef Island (Chapter 13), so this date may be later than is usual at Ramsay I.

of laying in the genus is retarded by about 6 days for every 1°C. lowering of April sea surface temperature. The relationship with temperature, rather than latitude, suggests that the main stimulus for breeding is likely to be connected with food supply, rather than with day-length. If we compare the mean surface temperatures for the month in which clutches are completed, we find that, for populations of Ancient and Japanese Murrelets listed in Table 4.1, all means

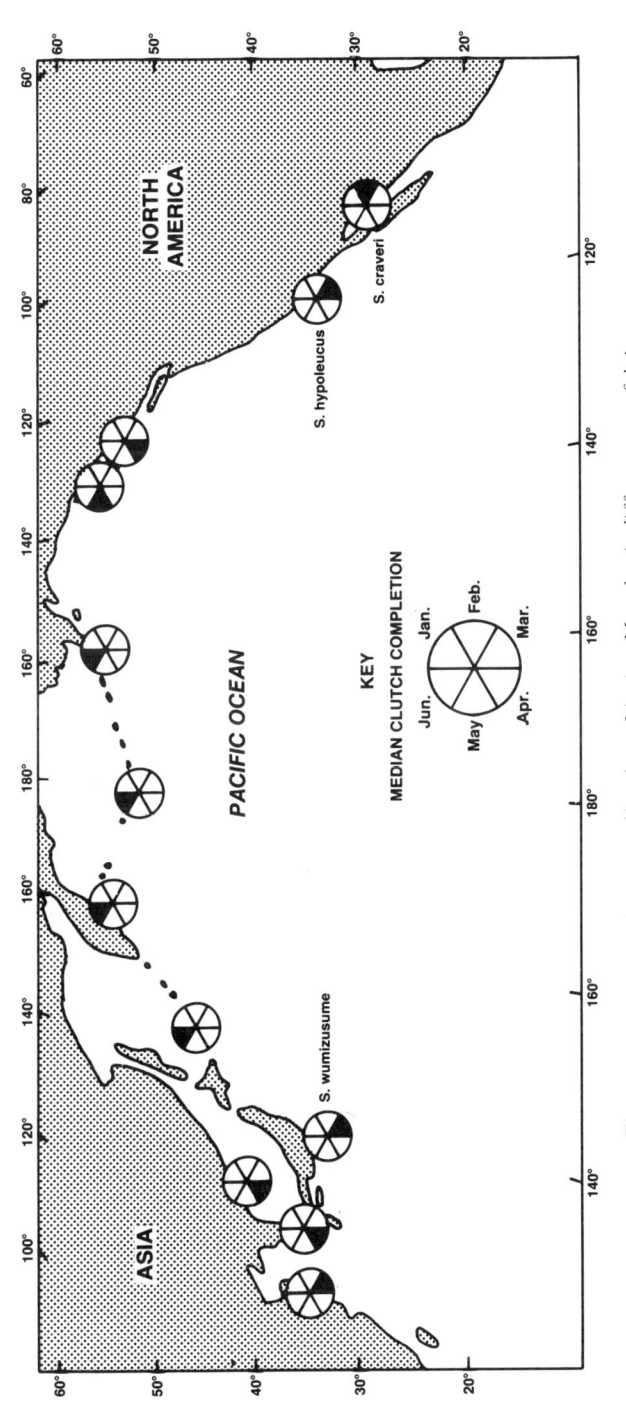

Fig. 4.1 Variation in the timing of breeding of Ancient Murrelets in different parts of their range.

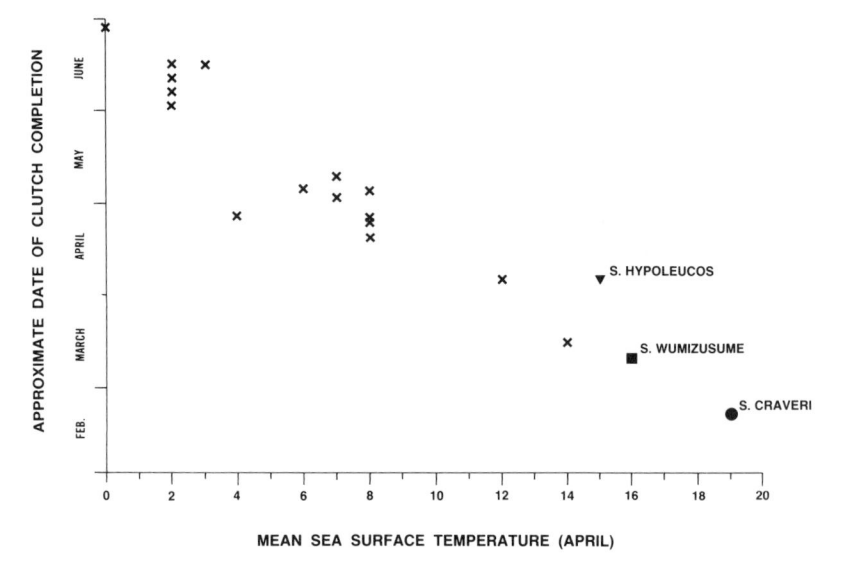

Fig. 4.2 *The relationship between the timing of egg-laying in Ancient Murrelets and sea-surface temperatures in the breeding area in April (temperatures from Anon., 1974).*

fall between 6 and 11°C. In California, Xantus' Murrelets lay in April, with a mean surface temperature of 13°C, while in Baja California, Craveri's Murrelets lay in February, with a mean surface temperature of 18°C. Hence, although the Ancient Murrelet restricts its breeding season to a fairly narrow range of sea surface temperatures, the two southerly murrelets breed in considerably warmer conditions.

There is usually a gap of 7–8 days between the laying of the first and second eggs, and then often another gap of 1 or 2 days before incubation is begun. Once initiated, incubation lasts about a month, if continuous. The chicks usually hatch within 12 hours of one another and they then leave on the second or third (rarely the fourth) night after hatching. The whole process, from the laying of the first egg, to the departure of the chicks, lasts about 42 days. However, the eggs are sometimes neglected for several days during incubation, extending the incubation period accordingly.

The parent Ancient Murrelets take equal shares in incubation, with most shifts lasting 2–4 days. Attendance at the colony by pre-laying birds continues to vary considerably from night to night, depending on the phase of the moon and weather and sky conditions, but birds arriving to exchange incubation seem relatively unaffected by weather, making their way to their burrows even on the stormiest nights (Jones et al., 1990). There is little information about whether breeders visit the colony except to take over incubation duty during this period, but it seems unlikely that they do so often.

Small numbers of non-breeding birds probably visit the colony throughout the breeding season, but there is a very marked increase in the rate of arrivals after about half way through incubation (at Reef Island usually about 10 May). As breeders begin leaving the colony, the nightly arrivals of non-breeders increasingly dominate the numbers present. By early June, after most breeders have left Reef Island, there are still large numbers of non-breeders visiting the colony, and numbers of birds seen on the gathering ground appear similar to, or larger than, numbers seen in the pre-laying period. There is much social activity on the gathering ground, including posture displays and vocalizations, and we assume that pair formation takes place then. At the same period there is also a lot of active burrow excavation going on and this presumably involves first-time breeders.

Activity at the colony begins to decline by about 3 weeks after the peak of family departures, with big nights for arrivals of non-breeders being spaced further apart, and numbers declining. We never stayed at Reef Island after late June, but anecdotal accounts from elsewhere in South Moresby suggest that little activity is seen after the end of the month. At Langara Island, Sealy (1976) observed sporadic attendance until early July. A generalized summary of the timing of events in the Ancient Murrelet's breeding cycle at Reef Island is shown in Figure 4.3.

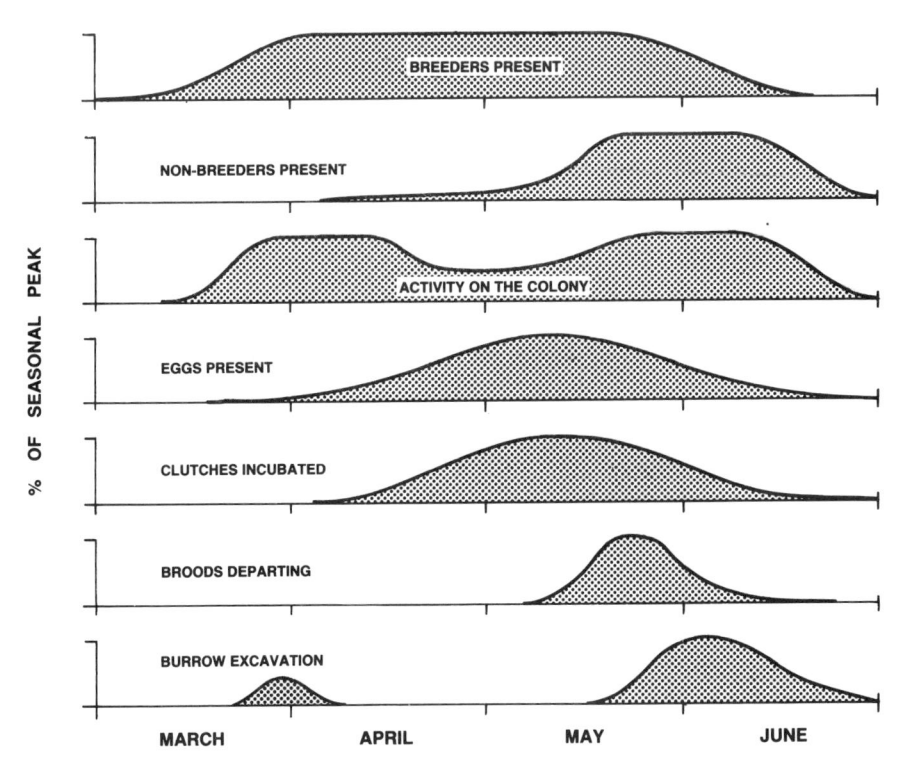

Fig. 4.3 *The timing of different events in the Ancient Murrelet's breeding cycle at Reef Island, British Columbia.*

The Ancient Murrelets' programme after departure from the colony is very poorly known. All observers agree that initial dispersal away from the colony is rapid and chicks are rarely seen near the colony during the period of departures. Litvinenko and Shibaev (1987) observed two family parties for about 5 weeks after departure and found that they remained in the same area, in inshore waters, throughout. The parents continued to feed their chicks, which were more or less fully grown by the end of the period. Elsewhere, Sealy (1975b) did not find family parties in inshore waters around Langara Island, but observed small numbers of solitary juveniles returning to inshore waters about a month after the peak of departures. No-one has observed family parties being fed in inshore waters around South Moresby, despite much time spent on the water by Mike Rodway, Moira Lemon and the Reef Island team. However, Guiguet (1953) saw several family parties throughout July around the Goose Islands, in the eastern part of Queen Charlotte Sound, where Ancient Murrelets do not breed. Similarly, Mike Rodway and Moira Lemon (pers. comm.) saw several family groups in June around the Moore Islands and nearby islands, along the mainland coast of Hecate Strait. They also saw seven family groups with young more than half grown in early July, close inshore along the northwest coast of Vancouver Island. It seems likely that these families originated from South Moresby, which is the nearest known breeding area. Their persistence suggests that these inshore areas may be regular brood-rearing areas, although the numbers involved seem to be fairly small. Most family parties probably remain out at sea while the chicks are dependent on their parents.

In British Columbia waters Ancient Murrelets more or less disappear in late summer and do not reappear until they begin to move south to winter quarters in October. Birds collected in October and November around Victoria, B.C., were all in very fresh plumage, indicating that they had moulted in the interim. By November, significant numbers are present in major wintering areas off California and Japan, but some remain within the breeding range, even as far north as St Lawrence Island in the Bering Sea (Bedard, 1966). Distribution patterns are fairly stable from November until February, when a general withdrawal from the winter range towards the breeding areas begins (Chapter 5). Stragglers may remain on their wintering areas as late as May, although a sample of birds collected in this month in Japan included a high proportion that were injured (Whitely, 1867). It seems that most of the population, breeders and non-breeders alike, move to the seas adjacent to the breeding grounds during the breeding season, unlike many auks, where young birds remain dispersed well outside the breeding range during their first summer (Brown, 1985). First-year birds occasionally visit the breeding colonies, but most apparently remain at sea (Chapter 14).

CHAPTER 5

The Ancient Murrelet at sea

Ancient Murrelets at sea; food and feeding behaviour, oceanographic affinities, non-breeding distributions, features of their marine habitat.

DIET

Compared to most marine birds, there is not much information available on the diet of Ancient Murrelets, probably because they do not bring food to their chicks at the colony. The only substantial studies of Ancient Murrelets diets were carried out around Langara Island, during the breeding season, and off Victoria, at the southern end of Vancouver Island, during winter (Sealy, 1975b; Carter and Sealy, pers. comm.). Around Langara Island, young Pacific sandlance and euphausid crustacea made up most of the diet (Figure 5.1). For the breeding adults, of which Sealy examined 61, euphausids comprised more than 90% of the diet, by volume, in April and May. In late March and early April these were mainly *Euphausia pacifica*, whereas later in the season the euphausids were entirely *Thysanoessa spinifera*. In June, fish, mainly juvenile sandlance, 30–60 mm long, made up about half of the diet. Subadults, of which 30 were collected, all after mid-May, ate approximately equal volumes of euphausids, and sandlance and smaller quantities of seaperch, with sandlance dominating the diet in July. Eight young of the year, collected between 10 July

ADULTS (BREEDERS)

SANDLANCE SEAPERCH EUPHAUSIA PACIFICA

THYSANOESSA SPINIFERA

SUBADULTS (NON-BREEDERS)

SEAPERCH EUPHAUSIA PACIFICA THYSANOESSA SPINIFERA

SANDLANCE DECAPOD LARVAE

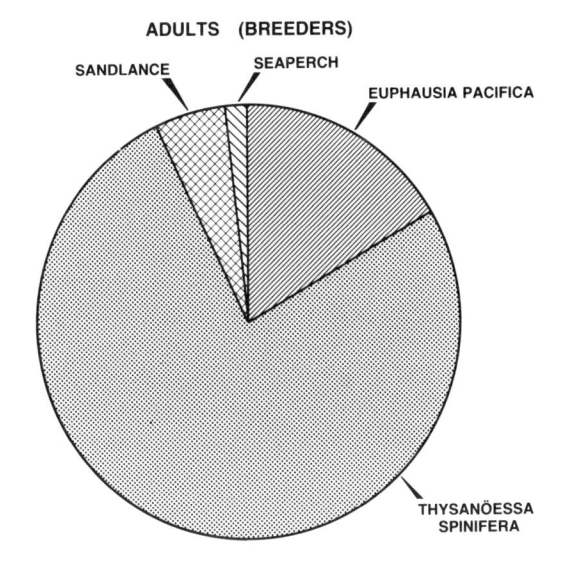

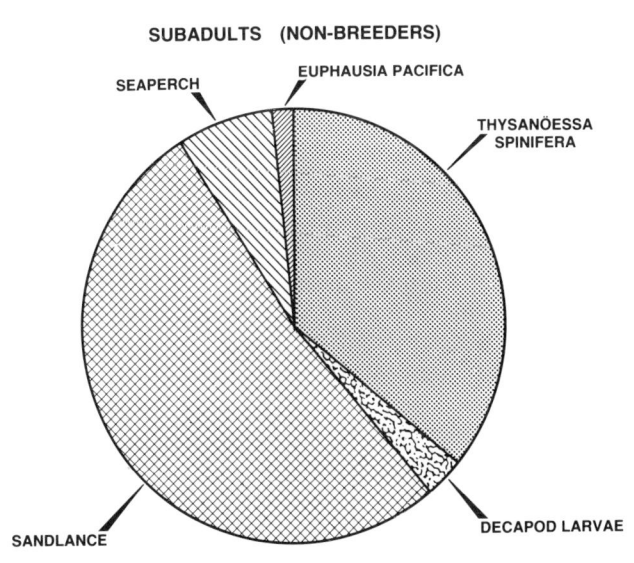

Fig. 5.1 *Diet of breeding and non-breeding Ancient Murrelets collected off Langara Island during March–June 1970 and 1971 (data from Sealy, 1975).*

and 4 August, and presumably just independent, contained practically nothing except sandlance over 30 mm in length.

Other information on summer diets supports Sealy's findings. Guiguet (1972) found euphausids to be the main food of the Ancient Murrelet in summer in British Columbia. Similarly, Piatt (pers. comm.), collecting

Ancient Murrelets around the Shumagin Islands, in the Gulf of Alaska, in June, found that most stomachs contained the euphausid *Thysanoessa inermis*. In the same area, nine Ancient Murrelets collected in late June and July (incubation period) contained mainly euphausids (Moe and Day, 1977). Stomachs of Ancient Murrelets taken by Peregrine Falcons in the Aleutian Islands in May and June contained, "principally the mysid (*Acanthomysis* sp.), a few euphausids (*Thyssanoessa longipes*), and remains of small fishes (*Ammodytes hexapterus*)" (White et al., 1973).

Vermeer et al. (1985) presented information on the stomach contents of 12 Ancient Murrelets collected in continental shelf waters off Graham Island in May, and nine collected in June. The earlier sample contained 56% euphausids by wet weight, the rest comprising sandlance, young rockfishes and other unidentified fish material. In June 25% of wet weight was rockfishes, and the rest unidentified fish. In the Gulf of Alaska, in summer, Sanger (1986, 1987) collected 18 Ancient Murrelets and found that they contained 55% euphasids (*T. inermis*) by volume and 42% fish, mostly capelin and walleye pollock. The general picture presented by these studies is that Ancient Murrelets feed mainly on invertebrates but take fish as a significant supplement. The same conclusion was reached by Hobson (1991), on the basis of an analysis of stable isotope ratios in Ancient Murrelet muscle tissues.

Young Ancient Murrelets remain dependent on their parents for more than a month after leaving the colony. During this time they are fed almost exclusively by their parents. In Peter the Great Bay, in the Sea of Japan, post-breeding Ancient Murrelets fed themselves and their young largely on larval herrings, although they also took amphipod and decapod crustacea and polychaete worms (Litvinenko and Shibaev, 1987). Near Buldir Island, in the western Aleutians, Alan Springer (pers. comm.) saw chicks about one-third grown being fed very small items, probably amphipod crustacea, by their parents.

In winter (Nov–Feb), Ancient Murrelets collected off the southern part of Vancouver Island were found to be feeding almost exclusively on *E. pacifica*, which was present in 83% of the 78 stomachs examined. Fish remains were present in 11% of stomachs, the only species identified being herring. Some stomachs contained very large numbers of *Euphausia*, the record being 1246 found in a female collected on 26 December. It is tempting to think that this may have been the murrelet's Christmas dinner! Many of the murrelets collected at this season were extremely fat, averaging much higher weights than we observed during the breeding season, at Reef Island. Weights were highest in January, when the maximum numbers of euphausids were found in the stomachs, suggesting that the peak weights coincided with a period when food was very abundant (Figure 5.2).

Aside from the collections described above, we have only anecdotal information on the diet of Ancient Murrelets. Chen Zhao-qing (1988) mentions, "fish, squid and shrimp", but the evidence on which the statement is

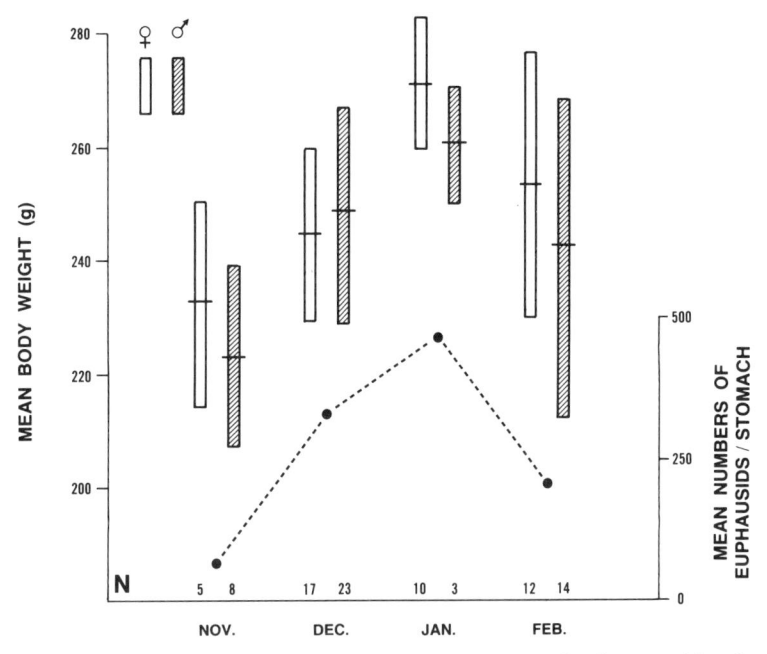

Fig. 5.2 *Changes in weight, and numbers of euphausids per stomach, of Ancient Murrelets collected off Victoria, British Columbia in 1978 and 1979 (data supplied by H. R. Carter and S. G. Sealy).*

based is unclear. Dementiev and Gladkov (1969) give the diet in the Commander Islands as, "small invertebrates, particularly gammerids". Most recent statements in general works on avian biology (e.g. Manuwal, 1984) probably derived their information from Sealy. I would summarize the summer diet as comprising large zooplankton, especially euphausids, and small (<10 cm) schooling fishes. Their diet in winter, known only from a single sampling area, cannot be generalized, but there seems no reason to think that it differs substantially from the summer diet.

MORPHOLOGY AND FEEDING ADAPTATIONS

In birds, the morphology of the bill and tongue have been widely regarded as being closely adapted to diet. Compared to the plankton-feeding auklets, the bill and tongue morphology of the Ancient Murrelet appears to be relatively unspecialized (Kuroda, 1954; Bedard, 1969a). The bill is fairly thin and pointed, although somewhat laterally compressed, and the tongue is narrow. The ratio of bill-width to gape is 0.197, similar to fish-feeders. The palette has much smaller denticles than those of the auklets. All these characters seem to place the Ancient Murrelet among the fish-feeding auks, while plankton feeders, such as the auklets, have bills which are broad and deep, and large denticles on the palette. Bedard (1969a: 195) comments that "as far as the morphology of the feeding apparatus is concerned, there is little doubt that they [murrelets] are true fish feeders."

Bedard's judgement was delivered at a time when almost nothing was known of the diet of the Ancient Murrelet, but in the same paper he suggested that "they [Ancient Murrelets] probably depend mainly on marine invertebrates ... worms, gammarids, molluscs and other small hard-bodied organisms probably make up a substantial part of their diet." (1969a: 190.)

So the unfortunate murrelets, handicapped with a fish-eating bill, must do their best to get along on the food that nature apparently did not intend. Perhaps this accounts for what Lewis and Sharpe (1987) described as their "rather spastic manner", although I am not sure exactly what was meant by that description. In any case, Ancient Murrelets can compensate in one way, because their skeleton appears much better adapted to swimming than those of the auklets. Kuroda (1954, 1967) placed the Ancient Murrelet second only to the murres in its physical adaptation for underwater pursuit.

Although information on the Ancient Murrelet's food is still very sparse, there is much more available now than when Bedard (1969a) discussed the morphology of auks in relation to their diet. Despite the apparently fish-eating characteristics of their bill, tongue and palate, it is clear that large zooplankton form an important, if not dominant share of the diet of the Ancient Murrelet.

Although the bill width/gape ratio of the Ancient Murrelet is similar to those of Xantus' and Craveri's Murrelets, the bill of the Ancient Murrelet is considerably deeper (7.2 mm versus 6.1 and 5.9 mm). This feature is shared by the Japanese Murrelet. In this respect, the Ancient and Japanese Murrelets are somewhat closer to the plankton-feeding auklets than are the other two *Synthliboramphus* species, and their deeper bills may indicate some concession to feeding on zooplankton.

With the benefit of much more information on auk diets, I think Bedard's generalizations can be modified, to some extent. The euphausid zooplankton, which form a prominent part of the Ancient Murrelet's food, are rather large—mostly more than 25 mm in length, in the case of birds examined by Sealy. They are about twice the length of the bird's bill, and hence bear the same relationship to the feeding apparatus as does the size of the average fish taken by the fish-eating murres (Bradstreet and Brown, 1985). Presumably the size of the prey, rather than its taxonomic affinities, determines the way in which it is handled, and this in turn determines the optimum shape of the bill. The diet of those auklets defined by Bedard as plankton eaters consists mainly of copepod crustacea, much smaller than the bill length (Bedard, 1969b). These may require a totally different method of handling, necessitating the high bill-width to gape ratio, and broad tongue characteristic of the group. Hence, I suggest that it is the relative size of the prey and predator that determines the morphology of the bill and mouth among auks, rather than the proportions of fish and zooplankton in the diet.

FEEDING BEHAVIOUR

Very little has been written on the feeding behaviour of Ancient Murrelets, probably because they usually feed out of sight of land. They appear to be gregarious while feeding. Sealy (1972) reported that they invariably occurred in parties of 6–20 birds when he saw them on their feeding grounds off Langara Island. However, 26% ($N = 162$) of feeding groups recorded by John Piatt (pers. comm.) around the Shumagin Islands of Alaska consisted of solitary birds, and a further 48% consisted of 2–6 birds. Only 15% of groups contained more than 20 Ancient Murrelets, although the largest group included more than 1000.

Austin and Kuroda (1953), describing the behaviour of the species off Korea, mentioned that flocks often dive in unison. Pairs of Japanese Murrelets also generally dive in unison (Austin and Kuroda, 1953; Higuchi, 1979). I gained a similar impression from those Ancient Murrelets that fed near Reef Island, when they occasionally did so. Parties of up to 18 birds swam roughly in line abreast, diving almost simultaneously, then popping up with less synchrony and more spread out, to regroup before diving again. On one day in early June, I was able to watch several groups feeding about 2 km from Reef Island, in water over 100 m deep. The average duration of 33 dives was 27 s (range 13–45 s), and the average of 29 intervals between dives, 16 s. These durations suggest that the murrelets may dive up to about 80 times per hour. Similar rates of diving, 73 and 53 times an hour, were recorded by Litvinenko and Shibaev (1987) for Ancient Murrelets feeding young on herring larvae. In both cases the dives were fairly short, compared with other auks (Burger, 1991). Assuming an average rate of descent of 1 m per second, the dive times suggest that the murrelets were not descending more than about 20 m. At Reef Island this was hardly surprising, because we had seen swarms of their favourite euphausids on the surface in the same area on many occasions. Harry Carter (pers. comm.) observed Ancient Murrelets off Victoria in winter scarcely submerging while feeding on euphausids, which mut have been within a few centimetres of the surface.

Around the Queen Charlotte Islands several species of seabirds are often found feeding together on concentrations of food. This phenomenon has been described in British Columbia by Sealy (1973a) and Porter and Sealy (1981). In 1971, 22% of 53 mixed feeding flocks seen off Langara Island contained Ancient Murrelets, with an average of 54 (range 1–200) in each flock. The most numerous constituent of these flocks was the Black-legged Kittiwake, while Rhinoceros Auklets, Glaucous-winged Gulls and Sooty Shearwaters were also common. Six Ancient Murrelets collected from the mixed flocks all contained the euphausid *T. spinifera* (Sealy, 1973a). Hoffman *et al.* (1981) found small numbers of Ancient Murrelets associated with large flocks of Sooty Shearwaters and Black-legged Kittiwakes feeding on capelin in the Gulf of Alaska.

We often saw similar gatherings to the north and east of Reef Island. The area to the east of Low Island seemed particularly attractive to seabirds and there, in April and May, we frequently found large flocks of Sooty Shearwaters and immature Black-legged Kittiwakes, always accompanied by smaller numbers of Rhinoceros Auklets, and sometimes by Ancient Murrelets and Cassin's Auklets. This area was also very attractive for large whales, particularly humpbacks. The smaller auks seemed inhibited by the presence of the other, larger birds, and were usually found on the fringes of the feeding throng, an observation also made by J. Piatt (pers. comm.) in Alaska. Large swarms of euphausids were often present in the area at the same time, and it seems very likely that these, or fish attracted by them, were the main cause of the seabird aggregations.

A very different feeding association involving Ancient Murrelets in Peter the Great Bay was described by Litvinenko and Shibaev (1987). There, the murrelets were feeding on schools of young herrings. The activities of the murrelets tended to force the fish to the surface, where Black-tailed Gulls gathered to feed on them. The gulls followed the activities of the murrelets closely, but it was the murrelets, rather than the gulls, which were in the centre of the flock. Flint and Golovkin (1990) gave a graphic description of an Ancient Murrelet attacking a shoal of young herrings, first herding it to make the fish clump together, and then dashing into the centre of the shoal. How this level of detail was observed is not clear.

Stejneger (1885) mentioned an unusual feeding habit that he observed among the Aleutian Islands. Ancient Murrelets dived repeatedly around the bows of his boat while it was underway, apparently making use of some turbulence created by its wash. Chase Littlejohn (in Bendire, 1895) mentioned similar behaviour:

> One would think at first that they were amusing themselves by flying a short distance ahead of the ship, dropping into the water and swimming in, so as to be near the bows as the vessel passed, thus diving beneath the hull and coming up again under the stern . . . they repeated this manoeuvre with unvarying precision throughout the entire day. By close watching I found that . . . they were feeding on small invertebrates such as are found on ships bottoms.

He mentioned that this behaviour continued until the ship reached shallow water close to shore. Although I passed through groups of feeding murrelets in the Queen Charlotte Islands while travelling in fishing boats, I never observed anything resembling this behaviour, nor have I found any reports from the 20th century. Perhaps modern anti-fouling paints have rendered the bottoms of ships less attractive to invertebrates and their murrelet predators.

NON-BREEDING RANGE

The Ancient Murrelet cannot be described as a migrant, in the sense of having clearly defined wintering and summering areas, but during the winter the population spreads well to the south of its breeding range (Figure 5.3). On the Asian side of the Pacific there are records as far south as Taiwan (Chang, 1980; Cheng, 1987) and Hong Kong (Chalmers, 1986). It is regarded as an accidental visitor in both places. Further north, large numbers occur in winter in the Sea of Japan, where the Ancient Murrelet was the commonest bird found dead from oiling in beached bird surveys carried out in winter in Niigata Prefecture, on the west coast of Japan (Kazama, 1971). It is also common in winter in the Strait of Korea (Seebohm, 1890; Gore and Pyong-Oh, 1971) and off the east coast of Honshu (Blakiston and Pryer, 1878; Kuroda, 1928; Yamashina, 1961).

In the northeast Pacific, moderate numbers occur south to Monterey (Ainley, 1976; Balz and Morejohn, 1977), small numbers off central California (Ainley, 1976; Briggs et al., 1987), and stragglers as far south as San Diego (Unitt, 1984) and Baja California (two records; Wilbur, 1987).

Despite these southerly movements, some birds remain throughout the winter in the most northerly parts of the species' range, with records from Adak Island, in the central Aleutians, in February (Byrd et al., 1974) and Kodiak Island, in the Gulf of Alaska, in November and February (Forsell and Gould, 1981). In the latter area Sanger (1972) found it abundant in February 1967. Cahn (in Murie 1959) reported Ancient Murrelets not uncommon around Unalaska Island, in the eastern Aleutians, during the winter. However, Zwiefelhofer and Forsell (1989) saw few Ancient Murrelets during winter surveys around northwest and southeast Kodiak Island, in 1979–84. Ancient Murrelets also appear to be unusual in the western Aleutians in winter, most arriving in April (Vernon Byrd, pers. comm.). Surveys in January and February of 1975 and 1976 did not record Ancient Murrelets in the Gulf of Alaska, although they were present off South-East Alaska and British Columbia, but by March and April they were common in the Gulf of Alaska, south of the Alaskan Peninsula and Kodiak Island (Gould, 1977; Harrison, 1977) and in the eastern Aleutians (Spindler, 1976).

On the Asian side of the Pacific, Stejneger (1887) saw small numbers of Ancient Murrelets off the Commander Islands in January and some birds winter among the Kuril Islands (Dementiev and Gladkov, 1969). Local eskimos told Bedard (1966) that a few Ancient Murrelets winter regularly in open leads in the pack ice off St. Lawrence Island, in the Bering Sea. Small numbers were seen from ships in April in the Navarin Basin, in the central Bering Sea (Gould 1977).

Southward movement does not seem to occur before October, when the first birds begin to arrive off Japan (Dementiev and Gladkov, 1969). At Mutsu Bay, in Honshu, where Kuroda (1928) kept year-round records, they started

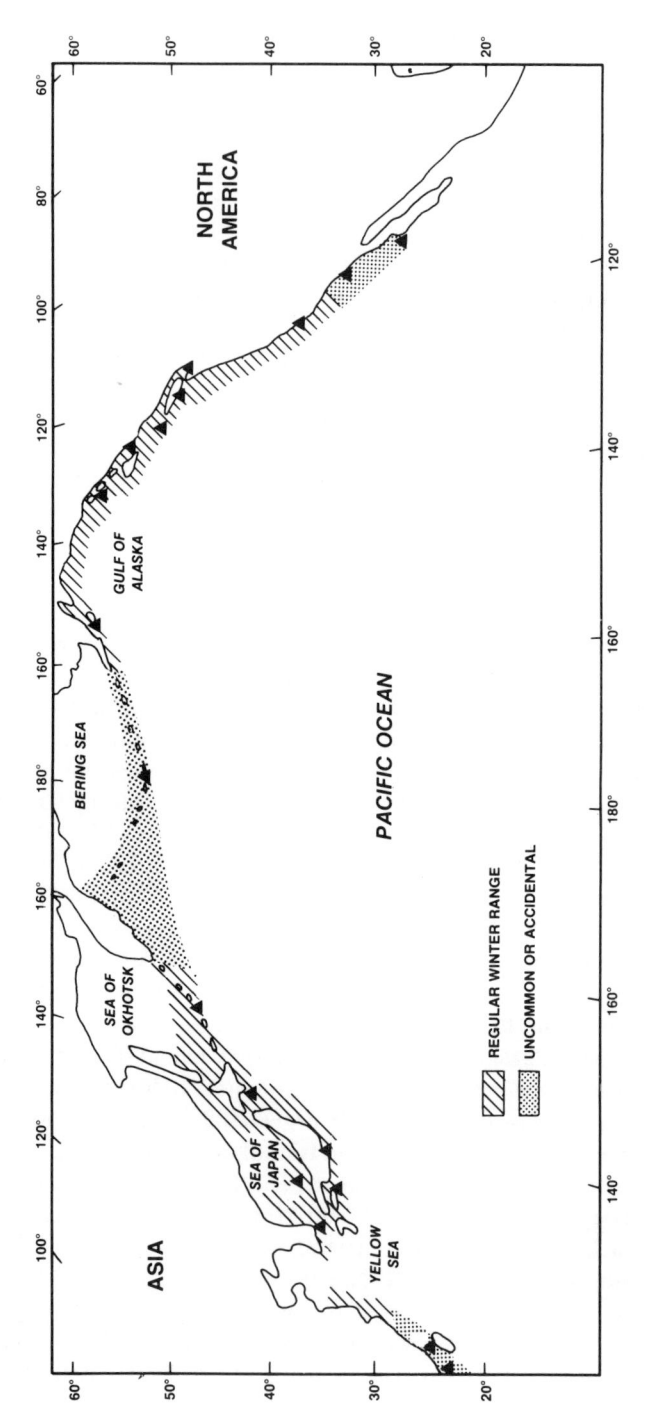

Fig. 5.3 *Winter distribution of the Ancient Murrelet.*

to appear in early November and were present until May. Austin and Kuroda (1953) reported them present off Japan mainly from November to May, Schrenk saw them of Southern Ussuriland during November to March (Taczanowski, 1893), and Gore and Pyong-Oh (1971) and Min and Won (1976) reported large numbers off Korea from December to February. Martins (1981) saw large numbers off Tokyo in late March and early April.

Immature birds begin to appear off the northern end of Vancouver Island in July (Martin and Myres, 1969). Offshore surveys at 49°N (the latitude of Barkley Sound) revealed small numbers from June to September (Vermeer *et al.*, 1987), mainly at the edge of the continental shelf. Intensive surveys along the West coast of Vancouver Island from May through to mid-October did not record Ancient Murrelets in inshore waters (Porter and Sealy, 1981). However, they are present in inshore waters at that latitude by November (Guiget, 1972; Campbell *et al.*, 1990). Regular surveys up to 120 km offshore from Westport, Washington during mid-April to mid-October produced only one record, of five birds 55 km offshore on 16 April (Wahl, 1975). In the Strait of Juan de Fuca some arrive by late September, but large numbers do not occur until mid-October (Lewis and Sharpe, 1987). By November, many are seen off Washington State, including Puget Sound (Slipp, 1942; Wahl *et al.*, 1981). Peak numbers there continue only until January, when large numbers may be driven ashore in Washington by gales (Balmer, 1936), but large numbers remain in the Strait of Juan de Fuca and Strait of Georgia into February, sometimes March (S. G. Sealy, quoted by Wahl *et al.*, 1981; Lewis and Sharp, 1987).

Ancient Murrelets arrive off northern California at about the same time that they appear off Japan, the first being seen in early October and larger numbers by late in the month (Ainley, 1976). Most birds have departed by the end of March, but a few remain as late as May. Maximum numbers of Ancient Murrelets are washed ashore in Monterey Bay during November to February (Stenzel *et al.*, 1988). In central California, south of Point Conception, the species was seen only from November to April, with most records in February to April (Ainley, 1976; Briggs *et al.*, 1987).

A small number of inland records of Ancient Murrelets, from interior British Columbia to the Great Lakes, have mainly occurred in late October and November (Fleming, 1912; Svihila, 1952; Guenther, 1965; Johnstone, 1964; Munyer, 1965; Smith, 1966). Hence, it appears that late fall may be a time of year when they are especially mobile. Munyer felt that most of these strays had been blown inland from the Pacific Coast by strong onshore winds, and their distribution fits with the timing of arrival of birds off Washington and California.

Some mystery surrounds the whereabouts of Ancient Murrelets during the late summer and early fall. Surveys carried out in Hecate Strait and off the west coast of the Queen Charlotte Islands in September failed to reveal any Ancient Murrelets, although they were present in small numbers on similar surveys in January and July and in large numbers from April to June (Vermeer and Rankin, 1984). As they have not yet reached their southern wintering areas by September, their absence from the waters around the Queen Charlotte Islands is hard to explain.

Early fall is the period when Ancient Murrelets moult and, in common with other auks, they presumably become flightless for some weeks at this time (Stresemann and Stresemann, 1966). Movements of some other auk populations during the post-breeding period tend to follow prevailing surface water currents (Brown, 1985), suggesting that they may allow themselves to be transported more or less passively at that time. As the offshore currents to the west of the Queen Charlotte Islands set northwestwards, it may not be too far fetched to think that the moulting murrelets are carried northwards into the Gulf of Alaska. Isleib and Kessel (1973) reported 1000 Ancient Murrelets, mostly in non-breeding plumage, in outer Prince William Sound in July–August, and several hundred in Yakutat Bay, well to the east of the main breeding areas of the species in the Gulf of Alaska. These birds could have been post-breeders from the Queen Charlottes. Small numbers of Ancient Murrelets have also been seen at the heads of major inlets along the south coast of Alaska in September (M. E. Isleib, pers. comm.).

A northward movement of Ancient Murrelets into the southern Bering Sea after breeding has been documented by several observers (Wahl, 1978; reports in Kessel, 1989). This may extend as far north as the southern Chukchi Sea, at least in some years, as there are records from that area in September, including sightings of small flocks flying northwards past Little Diomede Island on 22 September 1985 (R. H. Day in Kessel, 1989). These records, in an area where relatively few observations have been made at sea, suggest that a northwards dispersal during the late summer may be a regular part of the Ancient Murrelets schedule.

Unfortunately there is no information on the movements of Ancient Murrelets from banding. Although we had ringed more than 7000 adults and chicks at Reef Island by the end of 1989, by the end of 1990 only one recovery away from the Queen Charlotte Islands had been reported. This was of a bird ringed as a non-breeder at Reef Island in 1987 and found dead on the beach near Newport, Oregon the following winter. The finder reported that the corpse was very fresh, so presumably the bird died nearby, and this gives us a clue that some Ancient Murrelets from the Queen Charlotte Islands must winter south of British Columbia. However, without further evidence, it is hard to know whether it was a normal event, or just a rare straggler.

MARINE HABITAT

Accounts of the marine habitat of Ancient Murrelets are confusing and contradictory. While some authors hold that the species is largely offshore, and even pelagic, in its distribution, others imply that it is found mainly close to shore. For instance, Kuroda (1963) notes that the species was seen only within 20 km of land on winter surveys off northern Japan. Kozlova (1957) says that it is found, in summer, "in shallow, calm places, gulfs and bays", an opinion echoed by a more recent account of the species in the U.S.S.R. (Flint and Golovkin, 1990). Grinnell and Miller (1944) characterized the species as being found in, "inshore waters of the open ocean", off California, while Unitt (1984) noted that most records off San Diego County were of birds seen within 3 km of shore. The species is common in winter in the enclosed channels around the Strait of Juan de Fuca (Wahl et al., 1981), and in inshore waters off Victoria, where up to 6400 have been recorded on Christmas bird counts. In the Straits of Georgia, Ancient Murrelets are particularly associated with areas of fierce tidal currents, such as Active Pass, and Haro Strait, in winter (Campbell et al., 1990).

Baird et al. (1884, apparently based on observations by Dall) noted that the species frequented bays and harbours in the Aleutians more than other auks, and Forsell and Gould (1981) found them mainly in inshore waters in the Kodiak area in winter. In summer in the western Aleutians systematic surveys up to 300 km offshore showed that Ancient Murrelets were common (>1 bird/ km^2) up to about 30 km from land, but much less common further out, with only $0.05/km^2$ seen in the outermost survey zone (D. Forsell, pers. comm.).

In contrast, Briggs et al. (1987), reporting on a very extensive series of surveys made from boats and aircraft, off California, observed the species, "primarily seaward of the shelf break [edge of the continental shelf]", although only small numbers were involved. Balz and Morejohn (1977), operating off Monterey in November and December, showed similar results, finding Ancient Murrelets in pelagic waters, or less commonly, in continental shelf waters out of sight of land. Surprisingly, Sanger (1972), reporting on several cruises beyond the continental shelf off Oregon and Washington in October, November and January, never recorded the species at all.

Littlejohn (in Bent, 1919) reported sighting family parties up to 700 km from land, and Arnold (1948) observed them up to 160 km offshore from the Aleutians, in continental shelf waters, in June and July. Bartonek and Gibson (1972) found Ancient Murrelets to be the most numerous small auks in Bristol Bay, Alaska in July. All those seen were between 16–160 km from land, again in continental shelf waters. Their observations included four family parties, all with two chicks each. In an adjacent area, Troy and Johnson (1989) found Ancient Murrelets commoner in July than in other months offshore over the north Aleutian shelf.

During the breeding season Ancient Murrelets were rarely seen feeding within sight of land around Reef Island, although they occurred in large numbers 1–3 km offshore while gathering. In the early morning in May and June, flock after flock of murrelets could be seen flying off from the gathering ground eastwards into Hecate Strait, suggesting that they were doing most of their feeding well away from land. Similarly, feeding concentrations are rare within sight of other colonies in the Queen Charlotte Islands (M. Lemon, pers. comm.). At Langara Island during the breeding season, Sealy (1975b) found that murrelets fed occasionally within sight from the island, but that most fed well away from land. However, in July and August, some independent young of the year were found feeding in inshore waters. Systematic surveys in spring and summer around the Queen Charlotte Islands by Vermeer and Rankin (1984) and Vermeer *et al.* (1985) revealed large numbers in continental shelf waters, with highest densities near the shelf break. Their surveys did not extend beyond the continental margin.

Inevitably, if observers are mainly land-based, there is likely to be a bias towards sightings close to shore. However, most observations of Ancient Murrelets at sea have been made by ship-board observers. The majority of observations more than 50 km from land have been made in summer in the central north Pacific, or off California in winter. This includes birds caught in the gill-net fishery (Ainley *et al.*, 1981; Degange, 1983). On the basis of existing observations it appears that the species occurs mainly in continental shelf and slope waters in winter, especially off Japan and British Columbia, but extending beyond the continental slope in California. While breeding, it does not appear to leave continental shelf and slope waters. However, in summer, particularly in the post-breeding period, it may occur far offshore in truly oceanic conditions. It is possible that the disappearance of Ancient Murrelets from shelf waters around the Queen Charlotte Islands in late summer results from a movement out into deep waters beyond the continental margin, but this is pure speculation.

Part II
Studies at Reef Island

CHAPTER 6

Introduction

A brief description of the history and geography of the area, and of the marine environment. The role of the Ancient Murrelet in the ecology of the islands. Recent ecological change, and some comments on stability and diversity in ecosystems.

THE PACIFIC NORTHWEST

The Queen Charlotte Islands form part of a long, thin ribbon of moist coastal forest, squeezed between the Rocky Mountains and the Pacific Ocean, that stretches from northern California to the Alaska peninsula. The area enjoys a temperate climate with abundant, some would say superabundant, rainfall. Elsewhere in the Northern Hemisphere, in Europe, Japan and on the eastern seaboard of the United States, the temperate zone is heavily modified by human activities. In the Pacific Northwest, the natural ecosystems of the temperate zone, though already decimated south of British Columbia, take their last stand. It is worth considering why this is so.

When the merchant adventurers of the 15th and 16th centuries began to cover the globe in search of trade and plunder, they originated almost exclusively from Western Europe. From this base, in the days before the excavation of the Panama Canal, the northern Pacific was the most remote part of the globe. It was not until the 18th century that the tentacles of Spain, via its colony in California, of Russia, from the furthest boundaries of its Asian empire, and of Britain, interested in defining the western boundary of its

Canadian possessions, groped into the areas now known as Alaska and British Columbia (Pethick, 1980). One of the first things that they found was the sea otter, a small sea mammal, and its discovery caused a minor stampede (Ford, 1967). Sea-otters are members of the weasel family, which provides most of the world's most desirable furs: sable, mink, otter, ermine. Being entirely marine, and living in cold water, sea otters develop an even more luxurious fur than other mustelids. Such was its value, especially in the Chinese market, that it became worthwhile to maintain the enormously extended trade routes to the North Pacific.

European traders in the mid-18th century had a problem. Tea had just become the fashionable drink in Europe, but at that time it was available only in China. European trade goods of the period, mainly cloth and metalwork, held little interest for the Chinese, whose technology in those fields was well ahead of Europe. Consequently, traders had to pay for tea in gold, driving the price very high. When sea otter pelts, obtained at Nootka Sound, on the west coast of Vancouver Island, were exhibited in China by Captain Cook's crew, they found a ready market (Gough, 1980). The rush to the North Pacific was on. Much of the early exploration of the area was carried out by traders, such as George Dixon, who first circumnavigated the Queen Charlotte Islands ("C.L.", 1789).

Within a hundred years the sea otter had been virtually exterminated throughout its range, which stretched from the Commander Islands to California (Riedman and Estes, 1987). Other marine animals suffered similarly. Steller's sea-cow, found in the Commander Islands, disappeared before even a single specimen had been saved for science (Steller's own specimen, laboriously preserved while shipwrecked on Bering Island, had to be left behind because of lack of space on the boat built to carry him and his companions to safety). The great herds of fur seals that congregated in the Bering Sea were ruthlessly exploited. The decline of Alaska's fur resources was a major factor in disposing Russia to sell it to the United States.

Once the fur bearers had been struck down, the area offered little to encourage exploitation. Fish and forests were abundant, and mineral resources were good, but there were no markets, with only a handful of settlers on the west coast of the United States, and Japan a closed society. It was not until the 20th century, with the flood of people into California and the opening of Japan, that the natural resources of the Pacific Northwest began to be valued. From the second world war onwards, exploitation increased rapidly. Despite these developments, the coastal area between Vancouver, British Coumbia and Anchorage, Alaska remains one of the most sparsely settled and least developed parts of the entire temperate zone, and one of the few places where we can see natural temperate ecosystems almost untouched by human interference. Indeed, settlement has actually regressed in some areas, with the small communities that formerly supported local logging and mining industries replaced by outside workers commuting from southern cities.

Sunset over Louise Island, from Reef Island.

GEOGRAPHY OF THE QUEEN CHARLOTTE ISLANDS

The Queen Charlotte Islands form an elongated archipelago running parallel to the northern coast of British Columbia. They are separated from the mainland of North America by the trough of Hecate Strait, which narrows and shallows from south to north. In the north, Graham Island looks across the waters of Dixon Entrance to the outer islands of Alaska's Alexander Archipelago (Figure 6.1). The southernmost tip, Cape St. James, nearly 300 km to the southeast, guards the northern flank of Queen Charlotte Sound.

The archipelago comprises about 130 islands. The two largest are Graham Island in the north, and the deeply dissected Moresby Island, which forms a backbone to the southern part of the archipelago. These two are separated by the narrow waters of Skidegate Inlet, on the shores of which are the settlements of Queen Charlotte City, Skidegate and Sandspit.

A range of mountains, reaching 1200 m altitude in places, runs down the west coast of the entire archipelago, with a broad plain to the east, narrowing from north to south. To the south of Cumshewa Inlet, on Moresby Island, the land is mainly rugged and mountainous, riven by numerous deep, narrow inlets, and carved into a jigsaw-puzzle of interlocking islands. This is the area known as South Moresby, the Gwaii Haanas of the Haida, most of which has recently (1988) been set aside as a National Park reserve, pending the

70 *Studies at Reef Island*

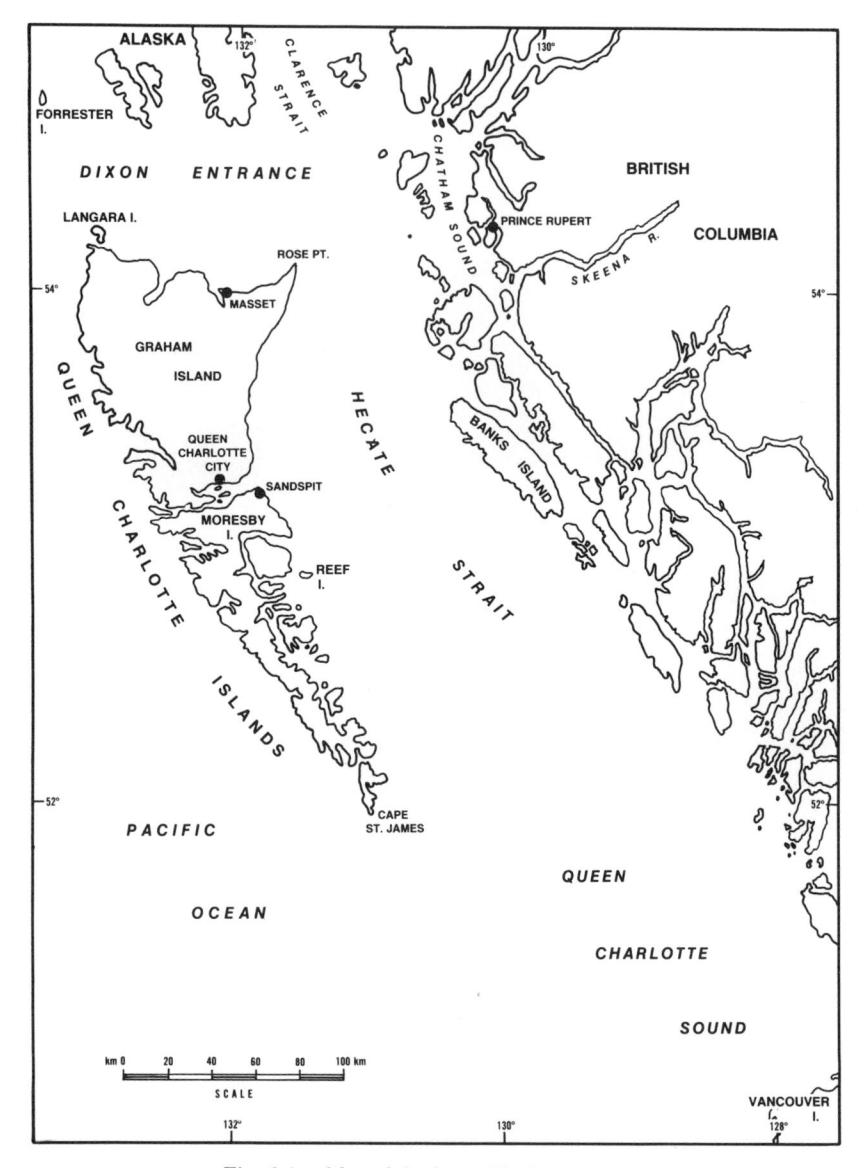

Fig. 6.1 *Map of the Queen Charlotte Islands.*

settlement of land claims with the Haida nation, who were the original inhabitants of the islands. Reef Island, where most of my own research has been carried out, lies just outside the northern edge of the park reserve, on the east side of Moresby Island. It has the distinction of being one of the most isolated of the Queen Charlotte archipelago, being further from the main islands than any other island of comparable size.

Lying between 52 and 55°N latitude, the Queen Charlotte Islands come under the influence of westerly winds throughout the year. These bring moist air from the Pacific, which provides copious year-round precipitation, mainly rain at sea level. The west coast receives about 250 cm of rainfall a year, while Reef Island, in the rain-shadow of the mountains, gets about 100 cm. Some snow falls in winter, but usually melts quickly. The climate is rather predictable; cool and oceanic, with minimum monthly temperatures varying from −5 to 12°C, and maxima from 3 to 18°C.

Storms are a regular feature of the islands, usually brought by major anticyclones originating in the Gulf of Alaska. In Hecate Strait the worst weather usually comes with an easterly, especially southeast, wind and periods of several continuous days of more than 60 km/h winds are common during the Ancient Murrelet's breeding season, from March to June.

Evening in the Misty Isles.

MARINE ENVIRONMENTS OF THE QUEEN CHARLOTTE ISLANDS

According to Kitano (1981) and Thomson (1989), the waters around the Queen Charlotte Islands fall in the North Pacific subarctic zone. There is a fairly steep gradient in winter sea surface temperatures between the northern and southern tips of the archipelago (Thomson, 1981), which may have some implications for the timing of breeding by the murrelets. To the north of British Columbia, the waters of the Gulf of Alaska form a gentle, counter-clockwise gyre, without any striking temperature gradients; a rather stable oceanographic regime. In contrast, the stronger California Current carries cold water southwards from the subarctic boundary, creating seasonal upwelling off California and driving the richly productive California marine ecosystem. The upwellings are subject to periodic failures due to the action of the El Niño southern oscillation (ENSO), and when these occur the large seabird populations of the Farallon Islands, off California, experience greatly reduced breeding success, owing to the dislocation of the local marine food web (Ainley and Boekelheide, 1990). In contrast, we have little evidence that the seabirds of the Queen Charlotte Islands, subject to more stable oceanographic conditions, are affected by periodic failures of such a magnitude. Nevertheless, conditions do fluctuate from year to year, as demonstrated by variations in the growth rates of Rhinoceros Auklet chicks at Lucy Island, on the mainland side of Hecate Strait (Bertram and Kaiser, 1988). Periodic failures have been recorded for some seabird species at Triangle Island, to the south of Queen Charlotte Sound (Vermeer et al., 1979; M. Rodway, pers. comm.). These fluctuations may be related to ENSO events, but the connection is less obvious than at the Farallons.

The Queen Charlotte Islands support a remarkable range of marine environments. On the broadest scale, the exposed west coast, with a narrow continental shelf, experiences very close to shore conditions typical of the open ocean. In Rennell Sound, on the west coast, albatrosses occur regularly within sight of land. In Hecate Strait, on the other hand, shallow banks extend far to the east of Sandspit and Rose Spit, providing habitat for inshore feeding birds well away from land. Birds such as scoters, Oldsquaws and Pigeon Guillemots may be seen sometimes almost in the middle of the strait.

Coastal environments are similarly diverse, ranging from the sandy shores of Graham Island, and the muddy estuaries at the heads of many inlets, to the rocky, high-energy shores that characterize much of the west coast and South Moresby. Masset Inlet, a long diverticulum of seawater reaching southwards into the centre of Graham Island, provides an excellent example of a brackish lagoon.

Although a full account of marine environments in the Queen Charlotte Islands is impossible here, there are a few characteristic phenomena which deserve mention. A very striking feature of the subtidal zone is the great development of kelp beds, which cover many hectares in Cumshewa Inlet, Laskeek Bay, and elsewhere. The trailing stems of *Macrocystis* may be more than 30 m long and, floating on the surface, they provide a constant hazard to outboard motors. These "forests" of kelp provide sanctuary for a wide variety of fish, particularly various rockfishes, and contribute very substantially to the primary productivity of the coastal seas. The extirpation of sea-otters from the archipelago (see below) has allowed the purple sea-urchin to become very numerous in places, and grazing by these animals, especially on rocky bottoms, has created "sea-urchin barrens" where algae are replaced by a carpet of spiny urchins. Despite this, there are still huge areas of kelp. In the

Queen Charlotte Islands, the principal effect of the urchins seems to have been to further diversify the available shallow water habitats, by providing numerous "clearings" where divers can observe an immense variety of subtidal life without having to hack their way through a jungle of kelp.

Steller's Sea Lions on the rocks to the southeast of Reef Island.

Marine mammals are very abundant around the Queen Charlotte Islands, although not necessarily more so than elsewhere in British Columbia waters. Steller's sealions are widespread among the islands, with the largest rookery in British Columbia being situated at Cape St. James, and a year-round haulout off Reef Island (Bigg, 1989). We have counted up to 450 at the Reef Island haulout. Harbour seals are also abundant and more widespread than sealions. They do not gather in such large numbers as sealions, but up to 20 were often present around Low Island. Elephant seals and northern fur seals are occasionally sighted.

Among whales, grey whales are annual visitors to Skidegate Inlet, where they arrive in early April and spend one or two months feeding before moving on northwards towards their summering grounds in the Beaufort Sea. Humpback whales and fin whales pass through Hecate Strait in May, and blue whales are sometimes seen, although I have never been lucky enough to do so myself. Killer whales are regular visitors, but those which occur in the islands are mainly transient pods without fixed territories. They seem to feed largely on other marine mammals, unlike the resident pods off Vancouver Island, which are mainly fish-eaters. Dall's porpoises and Pacific white-sided dolphins are common, but peripatetic.

Three fish support important fisheries in the Queen Charlotte Islands: salmon, halibut and herring. All the streams of any size in the archipelago support spawning salmon, many of several species, as well as cut-throat trout and steelhead. When the salmon are spawning, mainly in late summer, most mobile predators and scavengers, including many birds, converge to take advantage of the feast. Likewise, in April, when the herring are spawning, everything from gulls to whales moves inshore in their wake.

Herrings of pre-spawning age classes form an important food source for many birds throughout the summer. Shoals of immature fish occur frequently at the surface, where they often jump clear, making a calm sea suddenly erupt in a tiny "boil", as though assailed by a minute, but intense, downpour of rain. Such boils immediately attract gulls, kittiwakes and other mobile and opportunistic feeders. If the herring are well grown they also attract Bald Eagles, which occur in amazing density on the islands. I have seen as many as 60 eagles circling over a herring boil. In May it was not uncommon for more than a hundred to be seen on Low Island, just north of Reef Island, a favourite vantage point from which the eagles kept watch for potential food. Throughout these islands, the Bald Eagle is essentially a marine bird, feeding largely on marine fish and seabirds, including Ancient Murrelets.

Herring boils are often associated with swarms of small, shrimp-like crustacea of the family Euphausidae. This is the same family that contains the antarctic "krill" *Euphausia superba*. Euphausids provide food for the herrings, as well as for large whales, such as humpbacks and fin whales, and for several species of seabirds, especially Cassin's Auklets and Ancient Murrelets. The euphausids, in turn, live on copepod crustacea which are the first level of consumers, living directly on the primary producers, in the form of marine phytoplankton. Figure 6.2 shows a simplified food web for the waters around Reef Island. A complete diagram of the food webs of Hecate Strait would probably cover a baseball field at this scale, and would take several lifetimes of research to construct.

TERRESTRIAL ENVIRONMENTS

In a natural state, much of the Queen Charlotte archipelago, from a few metres above the top of the tide, to the subalpine barrens of the high mountains, was clothed in dense coniferous rainforest, dominated more or less equally by the western red cedar, the western hemlock and the Sitka spruce. The flat expanses of eastern Graham Island support extensive bogs, with scattered, stunted lodgepole pines. Flatly contradicting the idea of habitat specialization, the pines also dominate the few drier, rainshadow areas, particularly on south facing aspects and, in southeast Moresby, on poorly drained soils on exposed sites. Along valley bottoms, in swampy areas or where landslips have disturbed the ground, alders form dense stands. Other large trees

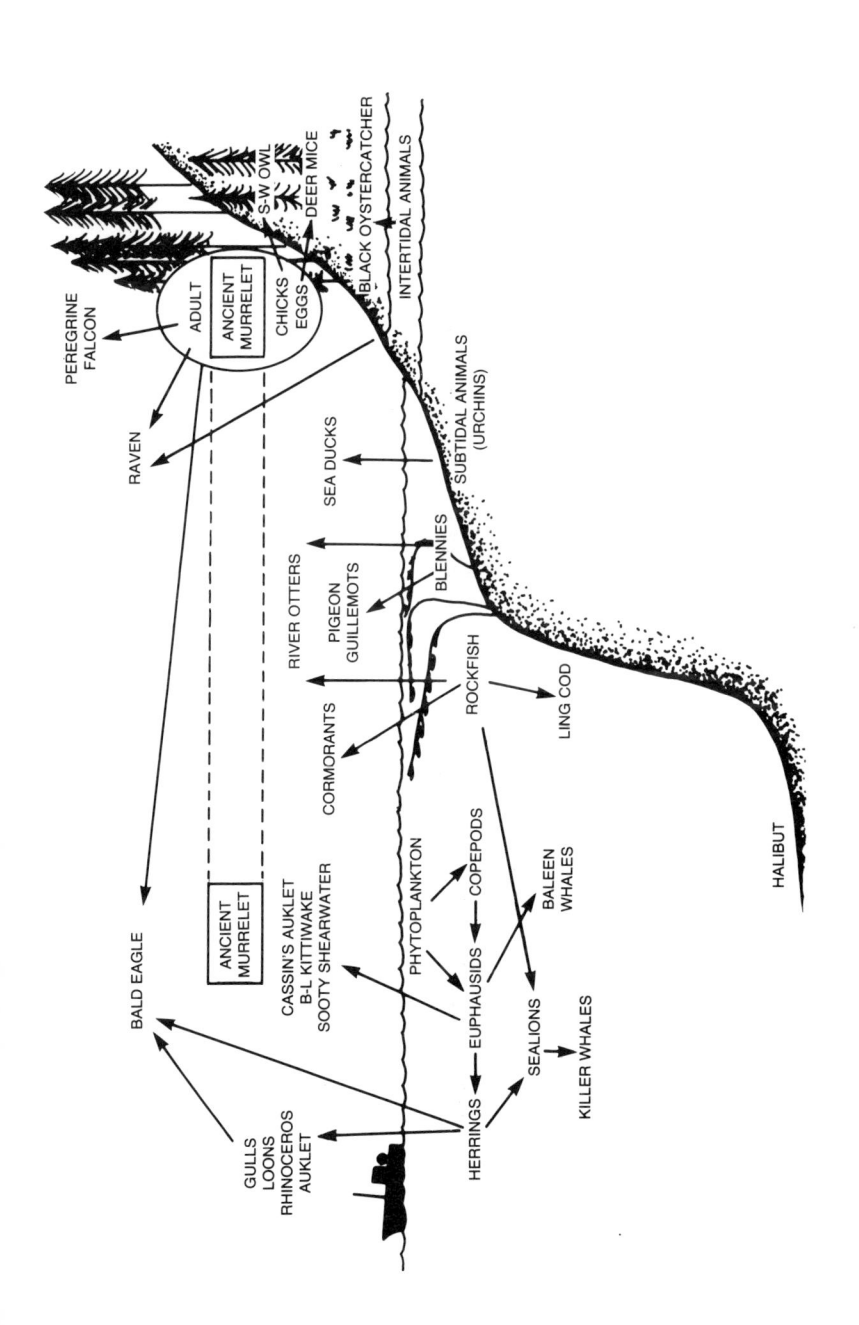

Fig. 6.2 The food web involving Ancient Murrelets in the Reef Island area.

Inside the mature forest, Reef Island.

(yellow cedar, yew) occur only in small numbers. A similar forest extends along the entire eastern rim of the Pacific, from Oregon and Washington States in the south, to the Alaska peninsula.

All of the three dominant conifers reach an immense size in lowland areas. At Windy Bay, on Lyell Island, in an area only saved by a whisker from the chainsaw, many Sitka spruce exceed 2.5 m in diameter at breast height. Trees more than 40 m in height are not uncommon. Even on Reef Island, where slopes are everywhere steep or precipitous, many old spruce and hemlock reach this size. With crowns generally closely packed, and many layers of branches, the undisturbed forest allows comparatively little light to reach the ground, so that the ground flora consists mainly of mosses and ferns, here and there enlivened by the pink blooms of the calypso orchid, the little waxy flowers of the single delight, or the ghostly white stems of indian pipes. In bogs and damp depressions the yellow spadix of the skunk cabbage, much loved by bears in spring, dots the ground.

Because of the activities of deer on most of the larger islands (see below) it is hard to judge what their shrub flora might have been like. On inaccessible cliff ledges, or on islands without deer, salal, huckleberry, salmonberry, and other shrubs form dense thickets. On Reef Island these shrubs are often found in little clumps on the upturned roots of fallen trees, or on top of large boulders, where they are out of reach of deer. In the absence of herbivores, the shrub layer might be much more extensive than we find today.

Where a blow-down has created an open space, regeneration of tree saplings is often so dense that it becomes impossible to walk through. The great depth of the canopy creates a torrential "rain" of leaves, twigs and branches that fuel the compost of the forest soil. The ground in most places, carpeted deeply in forest litter, is so soft that an arm can be thrust into it to the elbow without difficulty. Fallen tree trunks create "nurse logs" which sprout miniature forests of saplings. The impression in the interior of the forest is of being in the heart of an immense, silent machine working flat out, with primary production and decay chasing one another madly in a frenzied carousel of biological activity.

The resilience of this forest, in the face of the fierce maritime climate, is remarkable to anyone familiar with the sparse and stunted forests developed on the western coasts of Britain, a climatically similar area at the same latitude. In the Queen Charlotte Islands even small, steep islands, with shallow

An area of dense regenerating spruce on the site of a former windthrow.

soils, support dense, luxuriant forest, with trees more than 30 m in height. Where the forest is unable to maintain a foothold there are tangled thickets of salal shrubs. Only on the most exposed headlands are there the areas of the grassy turf so characteristic of oceanic islands elsewhere at similar latitudes. The conifers, especially the Sitka spruce, show a remarkable degree of salt tolerance, and in sheltered inlets it is not uncommon to find trees almost dipping their branches into the water at high tide.

Where trees and shrubs do not overwhelm them, usually on rocky headlands, or small islets, the Queen Charlottes can put on a splendid display of flowers. In May the Skedans Islands and parts of South Low and Low islands are washed by the blue of lupins, the yellow of monkey flower, and the magnificent drooping heads of red columbine. Chocolate lilies nod in the tall grass, and the rocks above high tide are dotted with yellow potentillas.

The almost complete forest mantle has forced those seabirds that wish to breed on the archipelago to choose between squeezing into the very limited open areas, found mainly on small islets, or on exposed headlands, or finding some way to utilize the forest. The Ancient Murrelet has taken the latter course and in the Queen Charlotte Islands breeds only in mature coniferous forest. Other seabirds also breed in the forest; Rhinoceros Auklets and stormpetrels commonly, and Cassin's Auklets occasionally. However, apart from the Ancient Murrelet, only the Marbled Murrelet, which nests high up in the trees rather than burrowing in the ground, breeds exclusively within the forest.

RECENT ECOLOGICAL CHANGE IN THE ISLANDS

When Europeans first arrived on the coast of British Columbia the Queen Charlotte Islands supported a population of some 7000 Haida Indians (G. M. Dawson, in Lillard, 1989). Their great cedar-built lodges clustered in permanent villages with populations in the hundreds at sites such as Ninstints, Tanu, and Skedans, while numerous lesser sites were occupied seasonally. The Haida maintained a thriving economy based exclusively on the harvesting of marine resources. One such resource, which attracted the immediate attention of European explorers, was the fur of the sea otter. As elsewhere in the Pacific Northwest, this commodity proved so attractive to the fur trade that within 50 years the species had been almost entirely extirpated from the Queen Charlotte Islands. The sea otter, although perhaps never very abundant nevertheless played an important role in the nearshore marine ecosystems of the region. It feeds heavily on sea-urchins, keeping their populations in check and thus, under certain conditions, preventing overgrazing of the kelp community (Estes and Vanblaricom, 1987). The disappearance of the sea otter from the Queen Charlotte Islands may have led to an increase in sea-urchin populations, and a corresponding decrease in the extent of kelp communities. Kelp beds contribute substantially to the productivity of nearshore waters (Duggins, 1983), and where they occur fishes tend to be more numerous and more diverse (Ebeling and Laur, 1987). Thus, at the very outset, the arrival of Europeans in the Queen Charlotte Islands probably created a significant change in the marine environment, one from which it has still not recovered.

Once the fur of the sea otter was no longer available, Europeans turned to the equally desirable fur of the northern fur seal. These animals breed on islands in the Bering Sea, but their travels during the non-breeding season bring them into pelagic waters off British Columbia, including those west of the Queen Charlottes. A thriving business in hunting fur seals, as well as whaling, was the main economic activity of the city of Victoria, during the 19th century (Busch, 1985; Murray, 1988). In the early part of the 20th century whaling factories were operated at Naden Harbour in the north of Graham Island and Rose Harbour in the south of the Queen Charlottes, taking a mixture of sperm whales and large rorquals, including the humpback whale (Hagelund, 1987). Commercial whaling continued from the Queen Charlotte Islands until 1942, and in Queen Charlotte Sound (from Coal Harbour on northern Vancouver Island) until 1967 (Webb, 1988). Unlike the otter hunters, the sealers and whalers did not succeed in extirpating their quarry, but that was because economic factors intervened. Both seals and whales are scarcer now than they were before the industrial world reached British Columbia (Busch, 1985; Evans, 1987), and this, like the disappearance of the sea-otter, must have altered the balance of marine ecosystems in ways that we cannot now appreciate.

Having left their mark on the sea, the Europeans proceeded to attack the land. Contact with the new arrivals rapidly decimated the Haida, and crushed their morale. A powerful, efficient and culturally sophisticated society was destroyed by smallpox and other infectious diseases within a few generations, leaving a sparsely inhabited vacuum. European settlers tried farming, and some still continues, but mining and fishing became the main activities in the 19th century. Then, as forests diminished elsewhere in the new world, the timber wealth of the Queen Charlotte Islands took over as their main economic attraction (Dalzell, 1968).

Commercial forest operations in the Queen Charlotte Islands began in the early 1900s and accelerated during the First World War, as the value of Sitka spruce for building aircraft was discovered (Dalzell, 1968). Early exploitation was mainly by shore logging, the timber being dragged directly to the beach by steam winches mounted on barges anchored offshore, and floated off. Practically all the prime timber within a few hundred metres of suitable shores was cut by this technique before 1950, much of it to supply the wartime aircraft industry. Felling of interior forests on the larger islands began in the 1930s and has intensified over the past 20 years as supplies of similar trees in accessible areas of mainland British Columbia diminished.

Some disastrous mistakes were made. Talunkwin Island, where steep slopes surround the fine natural anchorage of Thurston Harbour, was largely denuded of trees in the 1960s. Much of the island still shows little sign of recovery, the bare, eroded slopes mutely witnessing the improvidence of those who cut them (Islands Protection Society, 1984). The creation of the South Moresby/Gwaii Haanas National Park Reserve has eliminated the risk of further destruction of virgin forest in the area where most of the remainder survives. Elsewhere, much of the forest on the main islands has already been converted to timber rotations that make mature forests a thing of the past. Fortunately, we found no sign that Reef Island had ever been logged.

In a natural state the terrestrial mammal fauna of the Queen Charlotte Islands comprised only a few species: small deer mice of the genus *Peromyscus*, two species of shrews, the Dawson caribou, the river otter, ermine and pine marten, and the black bear (McTaggart-Cowan, 1989). The absence of the mink, a good swimmer and present on islands along the mainland side of Hecate Strait, is surprising. Despite the fact that the animals were farmed on Graham Island at one time, mink never became established in the wild. The black bears, which reach an exceptionally large size in the islands, are probably indigenous, although they are thought by some to have been brought from the mainland by the Haida. An endemic subspecies of toad completes the list of native quadrupeds (Islands Protection Society, 1984). Like Ireland, the islands have no snakes.

Just like their counterparts elsewhere in the globe, the European immigrants thought that they could improve on nature and began to augment the fauna of the islands. As usual, the introductions had unintended consequences, some of which could certainly have been predicted. When beavers were introduced in 1936 (Anon, 1936) they spread rapidly over the lowlands of Graham Island, converting bogs into lakes and drowning areas of forest in their usual fashion. They may have had some impact on the populations of the small number of indigenous freshwater fishes (Northcote *et al.*, 1989). Muskrats, which were introduced in 1924 (Pritchard, 1934), have also become established.

Less evident, but more insidious than logging in its effects on the vegetation of the islands, has been the introduction of the mainland Sitka black-tailed deer, which began at the end of the 19th century (Osgoode, 1901). In the absence of natural predators, these deer have multiplied to the point that the Queen Charlotte Islands have the most generous hunting season for deer of any place in Canada. Despite this extended hunting season, the deer are very common and, by the standards of the mainland, extremely tame. On Reef Island they frequently wandered through our camp, occasionally upsetting the garbage pail, or chewing someone's underwear hung out to dry.

The deer have also proved extraordinarily adept at crossing salt-water barriers, reaching practically all the islands large enough to support them. Many smaller islands, though not supporting permanent populations, show signs of periodic visits by deer. I had a graphic illustration of their powers of voluntary dispersal one day in June when I encountered a hind more than 1.5 km from the nearest land, swimming doggedly across Selwyn Inlet towards Talunkwin Island.

The deer have a considerable effect on the regeneration of western red cedar, so that forest companies have to go to considerable trouble to protect young cedars against deer browsing in the Queen Charlotte Islands (Pojar and Broadhead, 1984). They do not seem to have much affect on the regeneration of spruce or hemlock, which is normally rapid wherever slopes are not too steep, but their presence has a marked effect on the occurrence of understory shrubs such as salal and huckleberry. Where deer are abundant, which is almost everywhere that they occur, ground vegetation and shrubs tend to be sparse or absent. Where deer are absent, as on Low Island, close to Reef Island, the shrub layer can be completely impenetrable, particularly on southerly aspects where salal predominates. The changes brought about by the deer in the composition and structure of the vegetation may have some consequences for the Ancient Murrelet and for other seabirds. Rocky Mountain elk, which were also introduced, have been less successful than the deer, and are confined to part of Graham Island.

Another exotic mammal which is probably having a considerable impact on the archipelago's ecosystems is the raccoon. These were introduced to the main islands after the Second World War, at a time when raccoon fur was in some demand. They multiplied rapidly, becoming a pest almost instantly around Skidegate and Queen Charlotte City. In the South Moresby area they are continuing to spread southwards, living mainly along the shore, where they feed on crabs and other marine life in the intertidal zone. I have seen 20 animals at once foraging in small parties along a favoured stretch of shore. Their density is such that they seem almost certain to have a significant impact on the intertidal community, but nothing is known about their effects, as yet. They have crossed numerous salt-water barriers to appear on many of the smaller islands, and their presence has been implicated in the disappearance of seabirds from some small colonies (Rodway, 1991). Further damage to seabirds seems inevitable as their spread continues. Exterminating raccoons over an area as large as the Queen Charlotte Islands seems impossible. Consequently, an endless skirmish between seabird conservationists and the advancing army of raccoons seems inevitable. They have not yet reached Reef Island, but they are present, at least periodically, on the nearby Limestone Islands, which also support Ancient Murrelet colonies (L. Hartmann, pers. comm.).

The ship rat, that scourge of seabirds on many oceanic islands (Atkinson, 1985; King, 1986), has also found a foothold in the Queen Charlotte Islands. On Langara Island, it was probably responsible for the extirpation of several seabirds (M. Rodway, pers. comm.), and it has almost certainly decimated the breeding population of Ancient Murrelets there. Early reports suggested that most of the island within a few hundred metres of shore, was dense with their burrows (Beebe, 1960; Drent and Guiget, 1961), and even as late as 1977, when the Provincial Museum surveyed the island (B.C. Provincial Museum, unpubl. report), they occupied some areas which have since been deserted. Skulls and bones of adult murrelets found in many burrows confirm that the impact of rats was largely through killing breeding birds on the nest (Bertram, 1989). Rats have also reached the Ancient Murrelet colony at Dodge Point on Lyell Island, and the Tufted Puffin colony at Cape St. James. At the former site the colony area has contracted since the 1970s, while at the latter Cassin's Auklets and Tufted Puffins have decreased in numbers. Rats may be responsible in both cases (M. Rodway, pers. comm.). They are present on at least two other seabird islands, although their impact on those is unknown.

I have stressed the changes that have taken place in the Queen Charlotte Islands to counteract the impression, given by photographs of the magnificent forests of mature trees, that their ecosystems are virgin. In fact, changes as a result of human intervention are many and affect some key elements in both terrestrial and marine environments. A sad fact about many of these changes is that they took place before we had obtained any idea of how the ecosystems functioned under natural conditions. Consequently, we may never know exactly what the natural ecosystems were like, or how their constituent organisms related to one another. This is true for most of the rest of the globe, too, but the fact is particularly poignant when we remember how short a time it is since Europeans first became acquainted with the islands.

ROLE OF THE ANCIENT MURRELET IN THE ECOLOGY OF THE QUEEN CHARLOTTE ISLANDS

Although the Ancient Murrelet breeds from the coasts of China to those of British Columbia, hence spreading a quarter of the way round the globe, our present estimates suggest that about one third of the world's population breeds in the Queen Charlotte Islands of British Columbia. In this archipelago, it breeds exclusively in burrows beneath the canopy of dense, mature coastal rainforests. It is one of the most numerous seabirds breeding in these islands, and certainly the most widespread of those forming large colonies, being found on more than 30 islands (Chapter 3).

Although Ancient Murrelets are numerous, it seems unlikely that their numbers significantly affect those of the small fishes and large zooplankton that form their prey. These are taken by many other seabirds, as well as by larger fishes and whales. Hence the impact of the murrelets on the marine environment is probably small. However, during their breeding season, the murrelets form an important part of the diet of at least one predator, Peale's Peregrine Falcon (Beebe, 1960; Nelson and Myres, 1976). They may also be a locally important source of food for Bald Eagles and Ravens. Through their activities in burrowing among the roots of forest trees, they may make some contribution to the blow-downs that periodically flatten parts of the forest, opening the understory to colonization by shrubs and tree seedlings. In entering their burrows they frequently drag cones underground, perhaps planting the seeds accidentally in the soil. Their nest-holes provide excellent storage facilities for the large Queen Charlotte Islands deer mice, which, in turn, sometimes prey on unattended murrelet eggs and chicks. The impact of predation by deer mice, which fluctuate in numbers with the local crop of spruce cones, may vary from year to year, affecting the reproductive success of the murrelets. On islands where larger mammals are present (rats, raccoons, river otters) the adult murrelets may provide a seasonally important source of food for these predators. Hence, despite the fleeting nature of their visits to land, the murrelets are deeply embedded in the terrestrial food webs of their breeding islands.

DIVERSITY, STABILITY AND PRODUCTIVITY

Scientists want to generalize. It is not sufficient simply to describe nature. That is the job of the naturalist. To cross the boundary from natural history to science we should be able to use our observations to derive general principles. Among ecologists, it is a matter of widespread regret that, so far, our discipline has provided very few firm rules. However, if we have not yet formulated ecological laws, we have no lack of hypotheses.

It is a general observation that the number of species in a given community tends to be greater in warm climates than in cold ones. In particular, polar regions support very few species, while tropical regions support many. A general explanation for this difference in diversity holds that tropical latitudes, being climatically stable with very high productivity, allow the coexistence and coevolution of a great diversity of species (exemplified by the tropical rain forest), while the periodic very severe climatic events of the polar regions, and their relatively low primary productivity, prevent the establishment of any but a handful of organisms (e.g. the arctic tundra).

In temperate latitudes the situation is a little different. If we look at the terrestrial vegetation of the Queen Charlotte Islands, for example, we find a rainforest, with a very stable climate and (judging from tree growth) high productivity dominated by only a handful of trees, shrubs and flowering plants and with a very small suite of attendant terrestrial animals. However, if we look below the tideline, the situation appears to be reversed, with an enormous diversity of intertidal and subtidal plants and animals jostling one another for the available nutrients. Casting a line from the rocks of Reef Island you can catch a dozen different fishes, and turning over the stones on the beach, or checking the rock pools, you can discover a dozen more. Mussels, barnacles, limpets, chitons and abalone shoulder one another for hold on the rocks, while scores of different algae festoon them in browns, reds, yellows and greens.

We can provide special explanations for why the terrestrial ecosystems of the Queen Charlotte Islands should support few species, while the sea supports many; the widespread glaciation of the archipelago during the Pleistocene, and the lack of a land-bridge to the adjacent mainland, for example, probably go a long way to explain the poverty of the land. However, the fact remains, that while both the sea and the land experience stable climates and support highly productive ecosystems, the terrestrial ecosystem is strikingly poor in species, while the sea is strikingly rich. Ecology remains an inexact science. If my account of the Ancient Murrelet and its biology contains rather a lot of "probably" and "perhaps" then you will have to bear with me. As the Nobel laureate, Francois Jacob, said;

> "Science does not aim at reaching a complete and definitive explanation of the whole universe . . . It looks for partial and provisional answers for certain phenomena . . ."
>
> (Jacob 1982: 10)

CHAPTER 7

The island and the work

Why we chose Reef Island; a description of some of our methods, of the ecology and fauna of Reef Island, and a description of the terrestrial and marine birds.

HOW THE STUDY DEVELOPED

We selected Reef Island as a base for studying Ancient Murrelets because it was one of the closest colonies to a settlement (Sandspit, 40 km to the north), and because it appeared to support some fairly dense areas of burrows, where we could hope to study an adequate sample of nests in a fairly small area. Unfortunately, the north side of the island, where most of the Ancient Murrelet colony is situated, is very steep—precipitous in places. This made it difficult, sometimes hazardous, to move about in the dark, which is when we did most of our work, and hard to catch the birds, which found it easy to launch themselves into the air from the steep slopes.

Camp was established in the centre of the north side of the island, close to a moderately sheltered cove where we moored our inflatable boat. From there, we initiated a network of trails to obtain access to the steep slopes where the majority of the Ancient Murrelets bred. Although the island had been surveyed in 1983 by a Canadian Wildlife Service team, which had mapped the boundaries of the colony, we had no idea at the outset where the densest nesting areas were. Nor did we know much about the birds' behaviour. One of our original plans was to rig up red lights, powered by a generator and controlled by a dimmer switch, so that we could gradually illuminate an area of forest

The research station on Reef Island, 1985. It has since been demolished.

without disturbing the murrelets (so we thought!). However, as soon as even the faintest light became visible, the murrelets took off. The idea had to be abandoned. Plans for other manipulations of birds in burrows also foundered because the murrelets proved extremely sensitive to disturbance. We found that one insertion of a hand into an occupied nest-cup, even without removing the bird, caused 10% of pairs to desert (Gaston *et al.*, 1988). When incubating birds were removed in daylight, all subsequently deserted. We only did this a few times. In retrospect I should have read Spencer Sealy's (1976) paper a little more carefully at the outset, because his results showed that he had similar problems, as he later confirmed. The sensitivity of the breeding birds to any form of disturbance severely curtailed the type of observations and manipulations that we could attempt, and did much to determine the type of procedures that we eventually followed.

The research at Reef Island was conducted in two phases. In 1984 and 1985 we were concerned mainly with ways to assess the accuracy and repeatability of censuses that were being carried out by other members of the Canadian Wildlife Service. To do this, we monitored active and inactive burrows in a variety of ways. The results were published some time back (Gaston *et al.*, 1988a), and as they are mainly of technical interest, will not be repeated here. During the same period, Ian Jones carried out research on the behaviour of Ancient Murrelets, with particular emphasis on their vocalizations. We jointly trapped birds for banding and collected information on breeding biology to provide a background for our other studies.

In 1985 we developed a technique for mass-trapping of Ancient Murrelet chicks as they were leaving the colony, and from 1986 onwards, the emphasis of the project switched to banding, with the intention of estimating important population parameters such as survival and age at first breeding. In 1987 David Duncan carried out a post-doctoral research project at Reef Island on hatchling energy reserves in the auk family and became sufficiently intrigued with Ancient Murrelets to return the following year and radio-track family parties after their departure from the colony.

In 1987 and 1988 Alan Burger, of the University of Victoria, used our camp as a base for studies of feeding and diving in Cassin's Auklet. In 1989 Jean-Louis Martin, of the Centre National de Recherche Scientifique at Montpellier, France, also joined us to study the forest birds.

THE GEOGRAPHY AND VEGETATION OF REEF ISLAND

Dalzell (1968) described Reef Island as, "an untidy octopus, sprawling across Laskeek Bay". It consists of an elongated east–west ridge reaching 120 m above sea level, steep and in places precipitous on the north, and sloping more gradually to the south, before dropping precipitously to the sea. Several of the southern headlands and coves are flanked by cliffs and near the west end a big, sheer face cuts almost across the island. Two isolated stacks stand a little back from the shore on the south coast, one of which we dubbed "Cassin's Castle", because of the dense concentration of Cassin's Auklet burrows found on the top of it. Smaller islets and reefs are scattered off the south coast, which is indented by several deep bays (Figure 7.1).

In the area of the Ancient Murrelet colony, where we spent most of our time, the average slope of the ground was about 45°, being considerably steeper in places, and cut by ravines filled with unstable scree. These were not ideal conditions for wandering about in the dark, especially when slopes were slick with rain. In retrospect we may have been fortunate that, in accumulating some five person-years of nocturnal activities over the life of the project, we had only one accident requiring medical attention, and even then, no bones were broken. A little simple trail construction enhanced our ability to move about the island considerably, and reduced the erosion promoted by our frequent travels over trails connecting our principal study sites.

Most of the island is covered by a mature forest of spruce and hemlock, with smaller patches of red cedar, mainly near the crest of the steep, north-facing slopes. Several areas on the top of the island have suffered major blow-downs in the recent past, and these are covered by dense regeneration, making them all but impassable without a machete. The ground and shrub layers in the forest are sparse, although dense carpets of mosses occur in damp depressions. On southern slopes, particularly around the deep inlet that we called Capella Cove, there are unmixed stands of lodgepole pine. The forest canopy is more

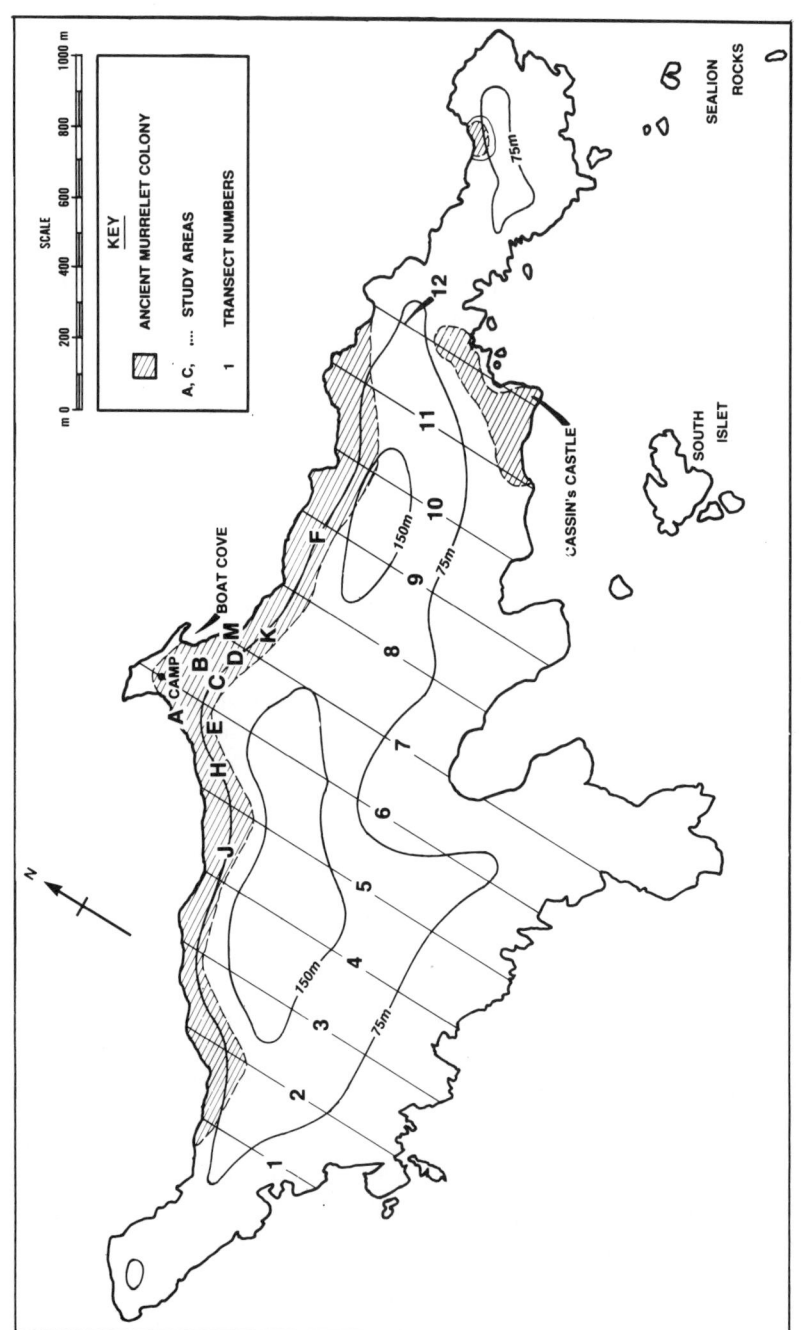

Fig. 7.1 *Map of Reef Island, showing the position of the Ancient Murrelet breeding area, and our census transects.*

open here, and the ground is covered in tall grasses, forming dense tussocks. Around the neck, at the east end of the island, and on some south-facing headlands, the forest gives way entirely to tussock grasses. Along the fringes of the north coast there are patches of red alder.

Near our camp site, in the middle of the north coast, there is evidence that the Haida had cleared a patch of forest to cultivate potatoes. The site was pointed out to me by Guujaaw, a visiting Haida, who identified the heaps of rocks which must have been cleared off the field. This clearing began to revert to forest about 40 years back, judging from the age of the spruce that has regenerated densely over the whole area. There is a midden containing shellfish remains nearby. The general enrichment of the spot is marked by a dense bed of nettles, the only place on the island where they occur. Dalzell (1968) notes that there was at one time a Haida fort on the island, but we found no evidence of it. Habitation was probably never more than sporadic.

Just beside the former clearing are several very large alders, and a group of huge, old spruce more than 40 m high, beneath which we built our cabin and pitched our tents when we began work in 1984. It is practically the only flat area on the island and, luckily, is adjacent to a tolerable anchorage, which we dubbed "Boat Cove". A small freshwater seepage nearby provided us, and the local deer and small forest birds, with water. Otherwise, surface water is confined to a few swampy pools on the top of the island, and to two small streams that empty into coves on the south coast.

The weather at Reef Island is very similar to that found along the whole of the East side of the South Moresby archipelago, being much drier than the western side. We kept weather records at camp, which we used to relate to variations in the behaviour of the murrelets. For comparison with weather elsewhere, a more standardized series, from a regular weather station at Sandspit airport 40 km north of Reef Island, is available (Figure 7.2).

THE FAUNA OF REEF ISLAND

Compared with the mainland of British Columbia, the Queen Charlotte Islands have a rather limited terrestrial fauna, as discussed in Chapter 6. Within the archipelago itself, and especially on the smaller and more isolated islands, the fauna is even less diverse. Among the land mammals of the Queen Charlotte Islands, Reef Island supports only deer mice, river otters, Sitka black-tailed deer, and little brown bats. We found the skull of a beaver, but it is hardly credible that beavers could survive on the island for any length of time. We saw no shrews, squirrels, martens or ermines. Nor did we encounter the archipelago's one and only reptile, the toad.

At sea, the situation is different. Small islets off the southeastern extremity of Reef Island are the site of a year-round haul-out for Steller's sealions. We counted up to 450 animals there, and numbers appeared to increase somewhat

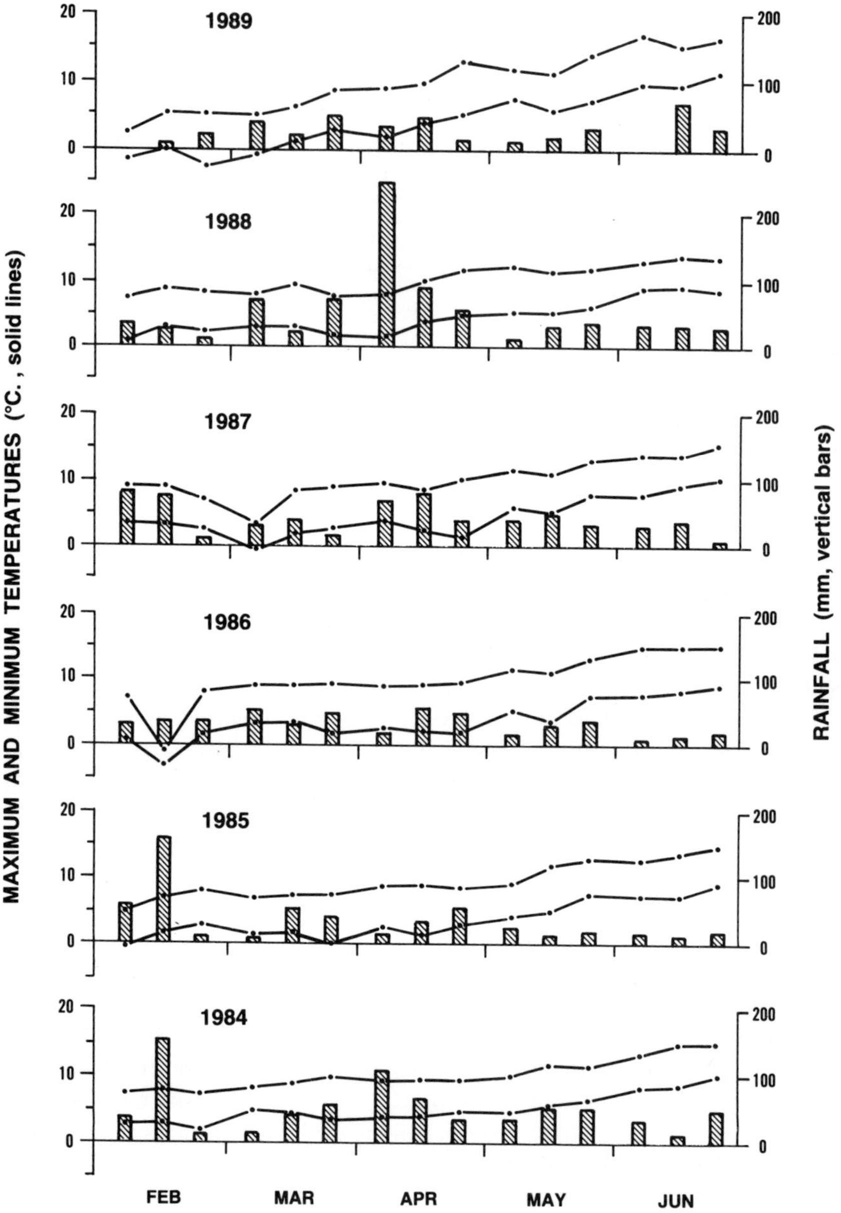

90 Studies at Reef Island

Fig. 7.2   *Temperature and rainfall records for Sandspit, British Columbia, during the periods of our study on Reef Island.*

during the period of the study. Sealions also hauled out periodically on the nearby Skedans Islands, where we saw up to 80. In the past, that haul-out may have been more regular. The slight increase seen at Reef Island may have resulted from some animals shifting away from Skedans. Harbour seals were also common around Reef Island, with up to 15 hauling out regularly on rocks below Cassin's Castle. A larger number was usually present on Low Island.

THE BIRDS OF REEF ISLAND

With its isolation restricting the arrival of mammals, Reef Island, like New Zealand, is largely a kingdom of birds. The forests of Reef Island support nearly all the birds characteristic of the Queen Charlotte Islands. Flocks of Red Crossbills and Pine Siskins call incessantly from the tops of the conifers. In April and May the cadences of Townsend's and Orange-crowned Warblers, the thin, sad note of the Varied Thrush, the liquid songs of the Hermit Thrush, the mosquito buzz of passing Rufous Hummingbirds and the calls of Golden-crowned Kinglets, Chestnut-backed Chickadees and Winter Wrens are heard everywhere. As the early singers reduce their volume in June, with their breeding efforts winding down, their place is taken by the ethereal cadence of Swainson's Thrush. Hairy Woodpeckers and Red-breasted Sapsuckers are common, while the fringes of the forest support Fox Sparrows and Song Sparrows. Of the species regularly encoutered in the forests of Moresby Island, only the Flicker and Steller's Jay were missing.

Among the larger birds, Reef Island supported two pairs of Ravens in all the years we worked there, one towards the west end of the island, and the other towards the east. Northwestern Crows were common. In the Queen Charlottes both of these corvids appear to feed predominantly in the intertidal zone, on shellfish, crabs and other marine life. The crows frequently fly out to isolated offshore reefs as the tide uncovers them, and on calm days they can even be seen picking their way gently across the floating festoons of kelp lying on the surface of the sea. Their habits would seem to bring them into competition with the avian shellfish specialist of the area, the Black Oystercatcher, although the two pairs of Oystercatchers which nested annually beside our camp appeared to take mainly limpets, judging from the shells that littered the nesting areas.

Four raptors occurred on Reef Island; Northern Saw-whet Owls, Bald Eagles, Sharp-shinned Hawks and Peregrine Falcons. The owls specialized in the local deer mice, although they also took a few Ancient Murrelet chicks. We seemed to hear their monotonous whistle in the Ancient Murrelet colony more frequently during the period when the chicks were departing than at other times of the year, suggesting that they deliberately sought out this seasonal food supply. Sharp-shinned Hawks may not have bred on the island every year, and we rarely saw them. However, in 1989 we located a plucking post which revealed that they fed mainly on Townsend's Warblers, with Song Sparrows and Brown Creepers also represented.

Two pairs of Peregrines attempted to breed on the island in all years, except in 1987 and 1989, when only one pair was present. At least five pairs of Bald Eagles were present in all years, although only two or three attempted nesting in most years, and only one, on the small islet off the south coast, was regularly successful in rearing young. Large numbers of non-breeding eagles, sometimes more than 100, frequented the island and nearby Low Island, particularly during May. This large concentration of non-breeders may have contributed to the apparently low breeding success of the resident pairs. The falcons, eagles and ravens were important predators of the murrelets and their ecology will be considered in more detail later.

. . . AND ADJACENT WATERS

Several species of seabirds, other than Ancient Murrelets, bred on Reef Island. Cassin's Auklets were the most numerous, with about 2000 breeding pairs scattered around the island in little pockets of up to 280 burrows (Rodway *et al.*, 1988). A dense group inhabited the summit of a small bluff on the south coast, which we named "Cassin's Castle". Like the Ancient Murrelet, the auklets visited the island only at night. They are semi-precocial, rearing their single chick to more or less adult size before it leaves the burrow. We investigated a small sample of burrows in 1989 and found that most eggs were laid in early April, with hatching taking place in the first 10 days of May. The majority of chicks departed in mid-June. Their diet at Reef Island was investigated by Alan Burger and David Powell (Burger and Powell, 1990), who found that the chicks were fed on euphausids and larval fish. By attaching dive guages to some of the breeders they were also able to discover that the auklets fed mostly at depths of 10–20 m (Burger and Powell, 1990).

Two species of Storm Petrels breed in the Queen Charlottes, the Fork-tailed and Leach's. Several hundred Fork-tailed Storm Petrels bred on Low Island, and we mist-netted them there several times, catching about 100 each night in a single 12 m net. Smaller numbers also bred on the little islet off the south coast of Reef Island, and in crevices around Cassin's Castle. We never had any evidence that Leach's Storm Petrels bred locally, although we trapped several while mist-netting for Fork-tailed Storm Petrels. Rodway *et al.* (1988) recorded Leach's Storm Petrels breeding on the Lost Islands, about 10 km south of Reef Island. Storm Petrels visit their breeding sites at night and we rarely saw them in daylight, except when severe southeast gales drove seabirds close in to Reef Island. Even then, we saw only small numbers.

Pelagic Cormorants bred on Reef Island in 1985 and 1986 when 13 and 11 nests were counted. Laying began in late May. In other years up to 300 cormorants, many of them immatures, and often including small numbers of

"Cassin's Castle" from the sea. Cassin's Auklet burrows were very dense on top of the small bluff in the centre.

Brandt's and Double-crested Cormorants, roosted on the rocky reefs off the south side of Reef Island, but none bred. All the cormorants appeared to feed mainly in shallow waters closer to the main islands, commuting regularly from their roost to Cumshewa Inlet and the inner reaches of Laskeek Bay. Another major roost site was located on Kingsway Rock, at the entrance to Selwyn Inlet. Although small, and apparently sporadic, the Pelagic Cormorant breeding site on Reef Island is one of only six in the South Moresby area (Rodway, 1988).

Glaucous-winged Gulls bred on Reef Island in small numbers each year, with several pairs nesting on a small islet close to the east end, and one or two more on adjacent parts of the main island. On Low Island there was a larger colony of about 40 pairs. In most years, eggs were laid in the second half of May, but in 1987, and again in 1989, there was little breeding activity and we suspected that the large numbers of Bald Eagles roosting on the colony probably prevented the gulls from laying. They may have done so after our departure in mid-June.

Far more spectacular than the seabirds breeding locally were the flocks of non-breeders present in the area for most of the season. The frequency of occurrence and the numbers involved fluctuated greatly from year to year. Some description of these fluctuations is worthwhile, because they can be compared to variations in the breeding regime of the murrelets. There also seems

Sooty Shearwaters at dusk close to Reef Island (Photograph by Ian L. Jones).

to be some correspondence between numbers of non-breeding seabirds and the occurrence of large whales in the area.

The commonest of the non-breeders were Sooty Shearwaters, transequatorial migrants from the southern Pacific Ocean. These were present throughout the periods when we were present on Reef Island, and appeared in large numbers whenever a major depression brought strong east or southeast winds to Hecate Strait. In 1985 and 1989 many thousands remained throughout May and June, and large feeding flocks were seen frequently, diving among euphausid swarms. Another common non-breeder, the Black-legged Kittiwakes also occurred in thousands in the same years. Kittiwakes seen in April were in breeding plumage, but by May the majority were immatures, the breeders having moved north to their colonies in Alaska. This species was not recorded at all in 1984. Blacklegged Kittiwakes occurred indiscriminately around euphausid swarms and fish shoals, and were often accompanied by immature Herring Gulls and smaller numbers of Rhinoceros Auklets, Glaucous-winged Gulls and Pacific Loons.

Year to year variations in the numbers of birds present seemed to affect species which feed partly or entirely on euphausids, such as the kittiwakes, shearwaters and Red-necked Phalaropes, more than the predominantly fish-eating loons and Rhinoceros Auklets (Figure 7.3). In 1985, and 1989, the big years for shearwaters and kittiwakes, we also had frequent sightings of Humpback Whales and a few of Fin Whales. These could easily have been taking advantage of abundant euphausids as well.

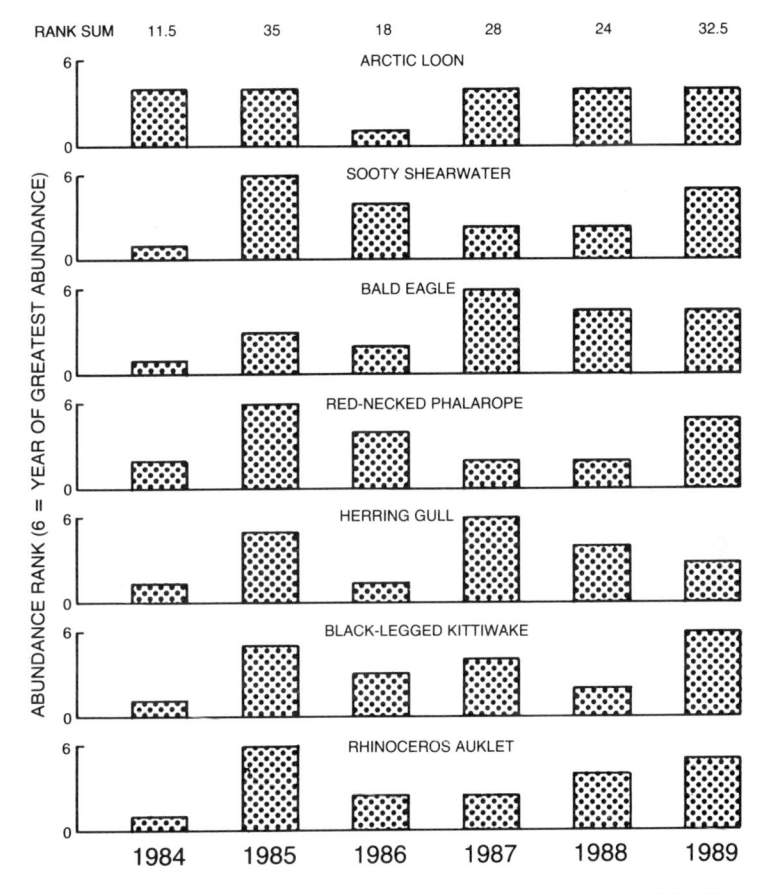

Fig. 7.3 *Relative abundance of marine birds seen commonly in the vicinity of Reef Island.*

METHODS

I shall describe many of the methods that we used when I deal with individual topics. However, information obtained from the trapping of adult birds and chicks, and from observations at study burrows, tends to be relevant to all sections of our work and therefore requires some description at the outset. Our methods evolved more or less continuously, as we found more efficient ways of obtaining information and as we learned about the effects that our activities had on the birds. However, there was a sharp discontinuity between what was done in 1984 and 1985, when we were fairly naive about the birds, and what we did later, once our first two seasons were evaluated.

To study breeding biology and attendance we selected several study plots within a half-hour's walk of camp. All burrows on these plots were investigated for evidence of breeding and access tunnels were dug to nest chambers, where

Seawatching from the lookout near camp; Louise Island in the background (Photograph by Ian L. Jones).

necessary. Each burrow was numbered with a small plastic label placed near the entrance. In 1984 we established five of these plots (A, D, E, F and H), but most of the burrows on plot A proved to be unoccupied, so it was dropped. We continued to check breeding success at the other four plots throughout the study, but plots D and F provided the largest samples, and most of the results presented will be based on these two only. Our methods varied over the years, as we became more concerned about the effects of our disturbance and strove to be less intrusive. In 1984 and 1985 we felt inside each nest chamber daily to check whether an egg had been laid, and subsequently, whether a bird was present. In 1986 and 1987 we cut this back to every three days during incubation, until close to hatching. In 1988 and 1989 we never put a hand into the nest chamber during incubation, and checked for the presence of an incubating bird by fixing a temperature probe in the nest cup before incubation had begun and then reading the temperature daily. Once the chicks had hatched, we banded them and the accompanying parent at the same time. We checked the next day to see if there had been an exchange of brooding duty, and if so we banded the other parent. However, we were able to catch both members of the pair at less than half the burrows. Only a few adults removed from the burrow at this stage of breeding deserted subsequently (3/221), although the small number that we removed during incubation often did so (Gaston *et al.*, 1988a).

Most adults were caught on the forest floor, with dip nets about 0.5 m in

Plastic tags placed in the mouth of a burrow.

diameter mounted on 2 m poles. In the early part of the night, catchers would sit quietly in a likely area, listening for the sound of birds landing. When one arrived, the catcher would spot the bird with a flashlight, then move quickly towards it, approaching from downslope as far as possible, and whip the net over it before it could rise, or just as it took off. Later in the night we patrolled the slopes gently to surprise birds sitting on the surface, but it was always easier to trap them just after arrival, when the non-breeders especially often seemed confused and disorientated.

Catching was easier on dark nights than on moonlit ones. When it was completely black some birds would fly towards our lights. Using a bright, quartz–halogen flashlight it was possible to attract birds from a distance under these conditions. A bird flying over, if held in the flashlight beam, would spiral down towards the light and land within a few metres. The technique worked particularly well on heavily overcast nights with light rain. However, even the small amount of light provided by starlight was sufficient to allow most birds to fly off before we reached them. When using this technique it was important to switch the light away from the bird if it began to fly off, because otherwise it often flew full-tilt into a tree. Presumably, departing birds headed for pale areas of sky among the black tree trunks. When illuminated with the flashlight, the relative brightness became reversed, with the trees appearing paler than the sky. The unfortunate birds then headed for the trees instead of the gaps. We never saw a bird injured by such a collision, but it did not seem a very good procedure. We trapped some birds in mist nets, but the length of time required to extract them in the dark made this a relatively ineffective method.

The area of the colony over which trapping was carried out was divided into zones and we recorded the zone in which each bird was captured or recaptured. In 1986 and 1987 a few birds trapped in burrows with chicks were fitted with radio transmitters to allow us to track them during and after the departure from the colony. We also fitted transmitters to a few breeders and non-breeders trapped on the surface.

An Ancient Murrelet with a radio transmitter glued to the feathers of the lower back.

The majority of birds captured on the surface were examined for the presence of brood-patches (the Ancient Murrelet has two lateral brood patches, each measuring about 20–25 mm across when fully developed), or eggs about to be laid (the egg makes a conspicuous bulge in the lower abdomen). The birds were weighed to ± 1 g on a Pesola spring balance, and we also took notes on the condition of their feathers (whether they were worn, whether there were any traces of winter plumage), and their feet, noting any damage to the webs. Any other unusual phenomena, such as muddy beaks or feet, or traces of blood on the plumage, were also noted. Because all this went on in pitch darkness, sometimes under stress (another bird lands beside you, it is raining, you are sliding towards the edge of a cliff, the batteries on your headlight are failing, etc.) the amount of additional information collected varied a lot according to the conditions.

At the beginning of egg-laying, in early April, the majority of females

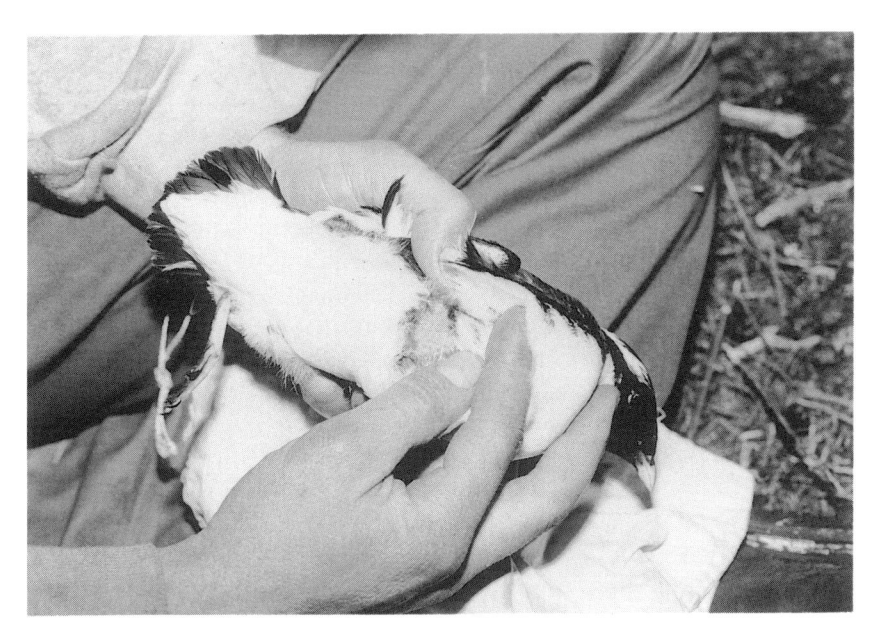

A fully developed brood-patch on an Ancient Murrelet.

trapped with oviduct eggs showed no signs of brood patches. By the end of egg laying, in the first half of May, the majority had a brood patch at least half-developed (Table 7.1). The interval between the laying of first and second eggs is 7–8 days (Chapter 12) and the development of the brood patch apparently takes place during this interval, so that by the start of incubation (usually 1–2 days after the laying of the second egg) most females probably have a more or less complete brood patch. Sealy (1976) reached a similar conclusion. The timing of brood patch development in the male is not known, but presumably it occurs at about the same time. Among birds with chicks, all but 1 out of 77 examined had fully developed brood patches, so the brood patch does not regress significantly before departure from the colony. Out of 310 birds banded with chicks, only one was trapped on the colony in the same year, after its brood had departed. The brood patch on this bird was partly refeathered,

Table 7.1 *Brood-patch development of females trapped with oviduct eggs (all years).*

Date	\multicolumn{3}{c}{Maximum diameter of brood patch (mm)}
	0	1–10	>10
Before 10 April	15	0	0
10–30 April	15	15	28
After 30 April	3	0	13

but was still clearly detectable. It appears that few breeders return to the colony after the successful completion of breeding.

Among birds trapped on the surface, the proportion with brood patches rose to about 50% by late April, then fell somewhat and thereafter hovered around 30–40% until the end of the season (Figure 7.4). Among birds trapped without a brood patch before 15 April, and retrapped after 12 May in the same year, 30/37 (81%) had fully developed brood patches when examined for the second time. Some of the birds retrapped without brood patches may have attempted to breed but never started to incubate, something that occurred fairly frequently (Chapter 12). Consequently, it appears that most birds present on the colony before mid-April will attempt to breed that year. Most of the birds trapped after the end of egg-laying (about mid-May in all years), either without brood-patches, or with brood-patches incompletely developed, were presumably non-breeders. However, a few may have been failed breeders, so our classification into breeders and non-breeders was not 100% accurate.

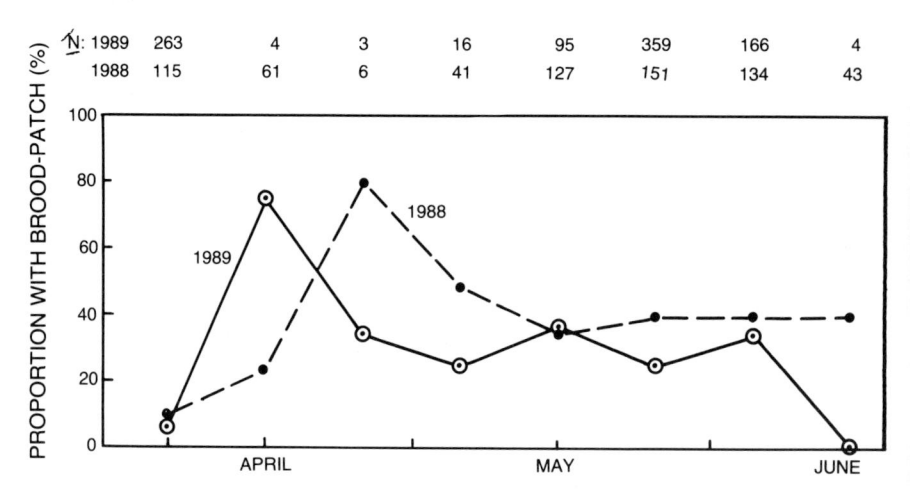

Fig. 7.4 *Proportion of birds trapped with brood-patches in relation to time of year.*

On the basis of the information obtained on brood patches I have treated the following categories of birds as breeders: those trapped on the colony before 15 April; those trapped in burrows, accompanied by chicks on the surface, or having oviduct eggs; and those trapped with a fully developed brood patch (>19 mm maximum diameter). I judged that birds trapped for the first time after 12 May in any year, and with a brood patch less than 11 mm across, were not going to breed in that year. Birds which did not meet the criteria for either breeders or non-breeders were not assigned a definite breeding status: these amounted to only 4% ($n = 1334$) of all birds not considered to be breeders.

From 1985 onwards, we trapped chicks on their way to the sea by erecting fences of clear plastic sheet about 0.5 m high, which directed chicks, always downhill, towards a small number of catching stations (up to four) close to the shore (Figure 7.5). The chicks were then banded and weighed as quickly as possible, and either released at the head of a cleared runway leading to the sea, or, if we had time, actually transported down the beach and released at the water's edge. The boulders of the beach presented significant obstacles to the chicks, so assisting them over this zone provided some compensation for any delay caused by diverting them into our funnels.

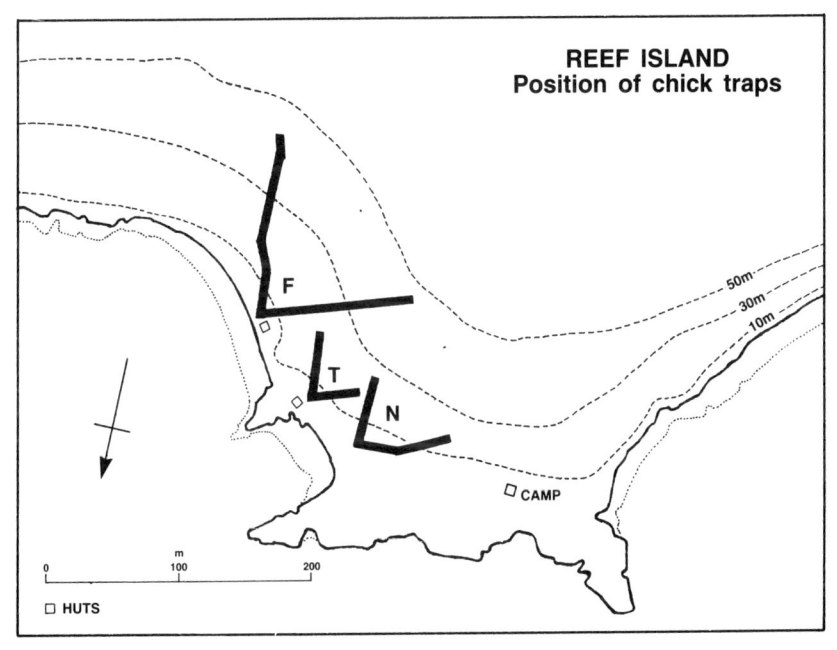

Fig. 7.5 *Position of chick trapping funnels used on Reef Island (F = Far, T = Tank, N = Near).*

From 1986 onwards we carried out trapping at the three regular stations (known as near, tank and far funnels) nightly from the start of departures until at least the first night at the end of the season when no chicks were trapped. A few chicks probably departed after we had ceased our operations, but we felt sure that we captured at least 95% of those that left the colony from the area above our catching funnels each year, providing us with a large and relatively unbiased sample of departure dates and weights.

MEASUREMENTS

At the start of the study, we felt that it was important to know something about the sex of the birds that we were dealing with. Consequently, Ian Jones and I collected a sample of 30 birds, dissected them, and measured them very

One of the chick-trapping funnels.

carefully to see whether it would be possible to distinguish the sexes on the basis of their external measurements. Although large series of measurements were available from museum specimens, the exact area of origin of many of those birds was in doubt, and it was likely that some shrinkage would have taken place after preservation. There was also the possibility that measurements might vary among different breeding colonies.

Among the birds that we dissected, males were significantly larger than

Table 7.2 *Measurements (in mm) of Ancient Murrelets at Reef Island. Sample sizes are given in parentheses.*

	Sexed by dissection					
	Male (15)		Female (15)		Females with eggs(30)	
Measurement	Mean	S.D.	Mean	S.D.	Mean	S.D
Wing	139.73	2.66	141.67	2.29	140.03	2.53
Culmen	13.47	0.69	13.36	0.38	13.40	0.56
Bill-depth*	7.22	0.22	6.67	0.21	6.68	0.38
Bill-width	5.25	0.26	5.22	0.23	5.59	0.32
Head and bill	59.55	1.24	59.01	0.82	59.06	1.48
Head-width*	22.44	0.32	21.72	0.55	21.93	1.01
Tarsus-length	27.22	0.73	26.92	0.56	26.74	0.95

* Difference between males and females significant ($P<0.01$).
Discriminant function: DF = (bill depth × 5.587) − (Bill width × 2.505) + (head width × 1.321) − (tarsus × 0.735) − 34.976.

females in the depth of their bills (measured at the level of the nostril), and the width of their head (measured at the broadest point, just behind the orbit) (Table 7.2). A discriminant function incorporating bill-depth, bill-width, head-width and tarsus-length, correctly sexed 26 out of 30 females trapped with oviduct eggs (Jones, 1985). However, the procedure of accurately measuring the birds in the field, in the middle of the night, took quite an appreciable time, reducing the number that we could capture. Also, with a regular turn-over of personnel it became hard to maintain uniform techniques for taking measurements. After 1985 we were keen to increase the number of birds trapped, so we reduced the measurements that we took to the one that differed most between the sexes, bill-depth. This also has the advantage of being fairly easy to take. Although about 30% of birds fall in the overlap zone for this measurement, I have used it as an indication of sex where a group of birds trapped in a particular situation, or activity, proved to have a mean bill-depth close to the mean for one sex or the other.

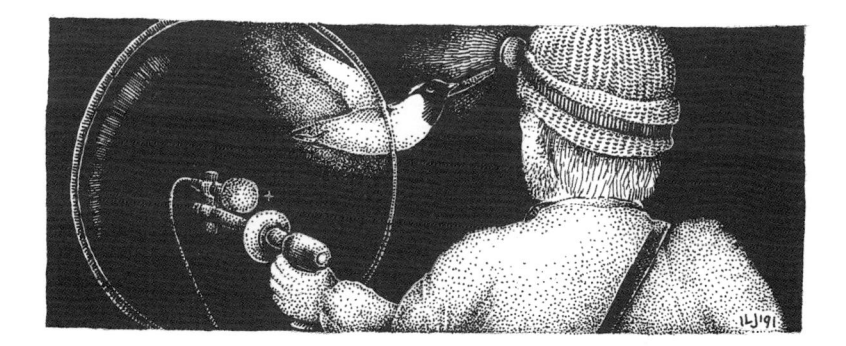

CHAPTER 8

Attendance and behaviour at the colony

Behaviour during the pre-laying, laying, and incubation periods; the departure of family parties; how the chicks find their way to the sea, and how they find their parents again; the first days at sea.

VOICE

The Ancient Murrelet has a wide variety of calls, as befits a bird which does a lot of its socializing in the dark. I shall describe them briefly at the outset, so that they can be referred to later, in the context of the general behaviour of the birds. They have been investigated in detail by Ian Jones, and most of my description is taken from the results of his research (Jones, 1985; Jones *et al.*, 1987a, b, 1989). He also kindly supplied many of the accompanying sonagrams.

Ian identified nine different calls given by Ancient Murrelets. The most common, the *chirrup*, is a short, emphatic, rather sparrow-like call, lasting less than half a second. On a sonagram, it can be seen to comprise a number of chevron-shaped tracings, rapidly repeated, in a distinctive array. When we inspected spectrograms from several Ancient Murrelets, it was possible to recognize those of different individuals on the basis of the number and relative size of these chevrons (Figure 8.1). Measurements taken from the spectrograms showed that the differences among the *chirrups* given by different indi-

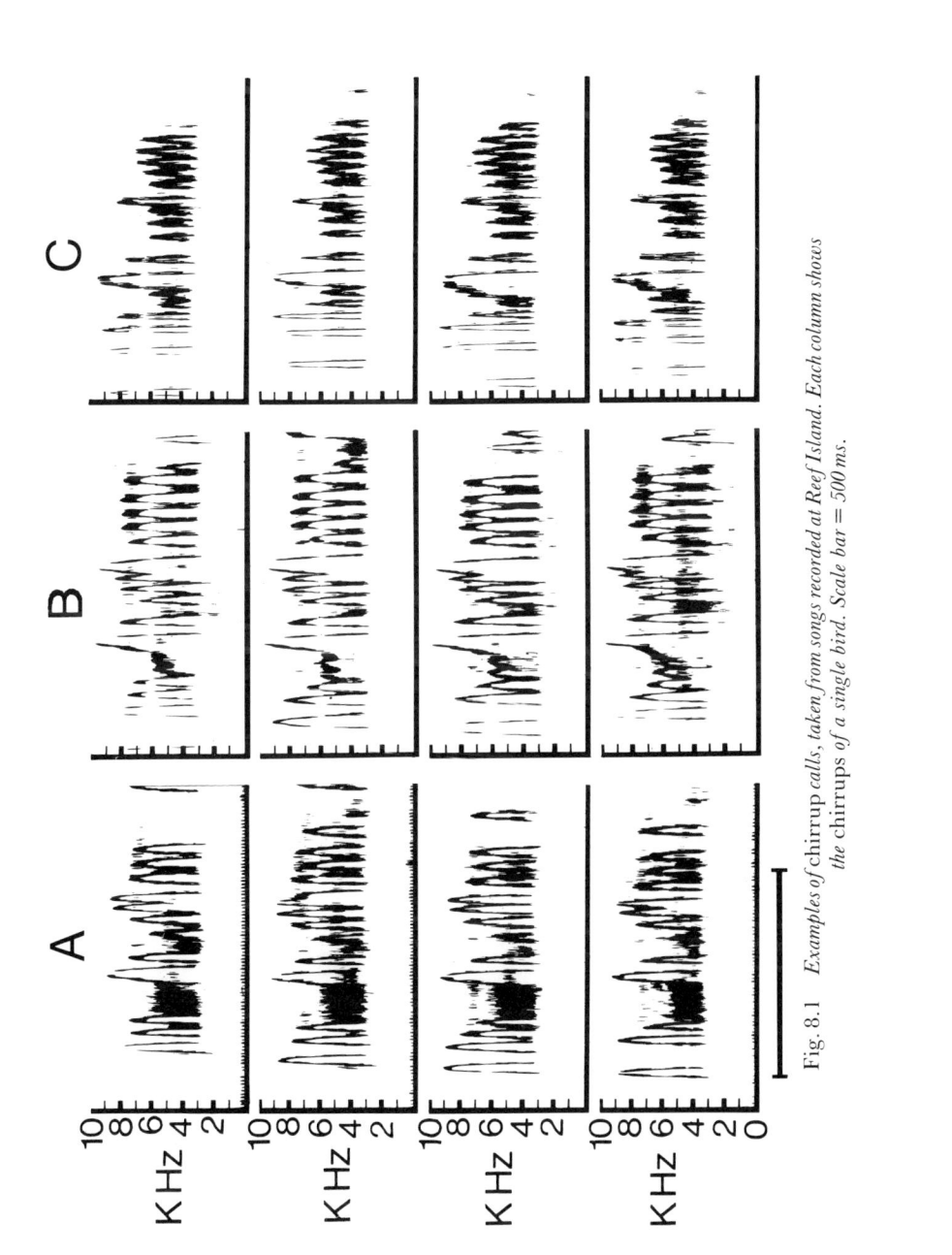

Fig. 8.1 Examples of chirrup calls, taken from songs recorded at Reef Island. Each column shows the chirrups of a single bird. Scale bar = 500 ms.

106 Studies at Reef Island

viduals were significantly greater than those among the *chirrups* of a single bird (Jones *et al.*, 1989), showing that each individual has a unique call.

Elements very similar to those which comprised the *chirrup* also occurred in several other calls; the *chip*, the *bubble*, the *chatter* and the *trill-rattle*. The *chip* was a single, sharp note, shown on a spectrogram as a single chevron, while the *bubble* consisted of an irregular series of chip-like notes, lasting up to several seconds. The *trill-rattle* was a brief burst of chip notes, delivered very rapidly at evenly spaced intervals, and lasting less than half a second. Finally, the *chatter* was a prolonged, high intensity burst of calling, which included elements of all the other calls mixed up in variable ways, and including loud rasping notes (Figure 8.2). It seemed to be given by birds which were highly aroused.

The *song* consisted of one, or more, chirrups, connected by well-spaced "chip" notes, and forming a rhythmic pattern (Figure 8.3). Like the *chirrup*, the *songs* of different birds could be recognized on sonagrams, and also often by an experienced listener. Certain individuals could be recognized from night to night, and even, occasionally, from year to year, on the basis of their distinctive

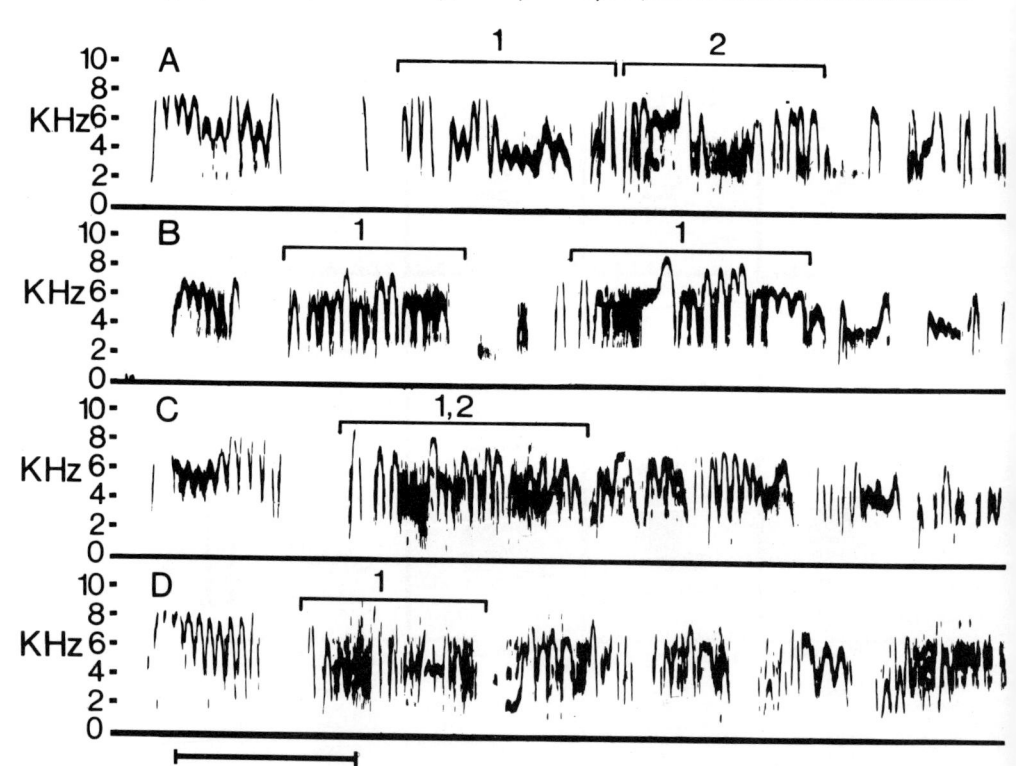

Fig. 8.2 *Examples of* chatter *calling recorded during the first 2.5 s of greetings interactions, recorded during incubation at four different burrows on Reef Island. A includes* chirrup *calls by both birds, B contains two* chirrups *of the same bird, while C contains* chirrups *by both birds, overlapping. Scale bar = 500 ms.*

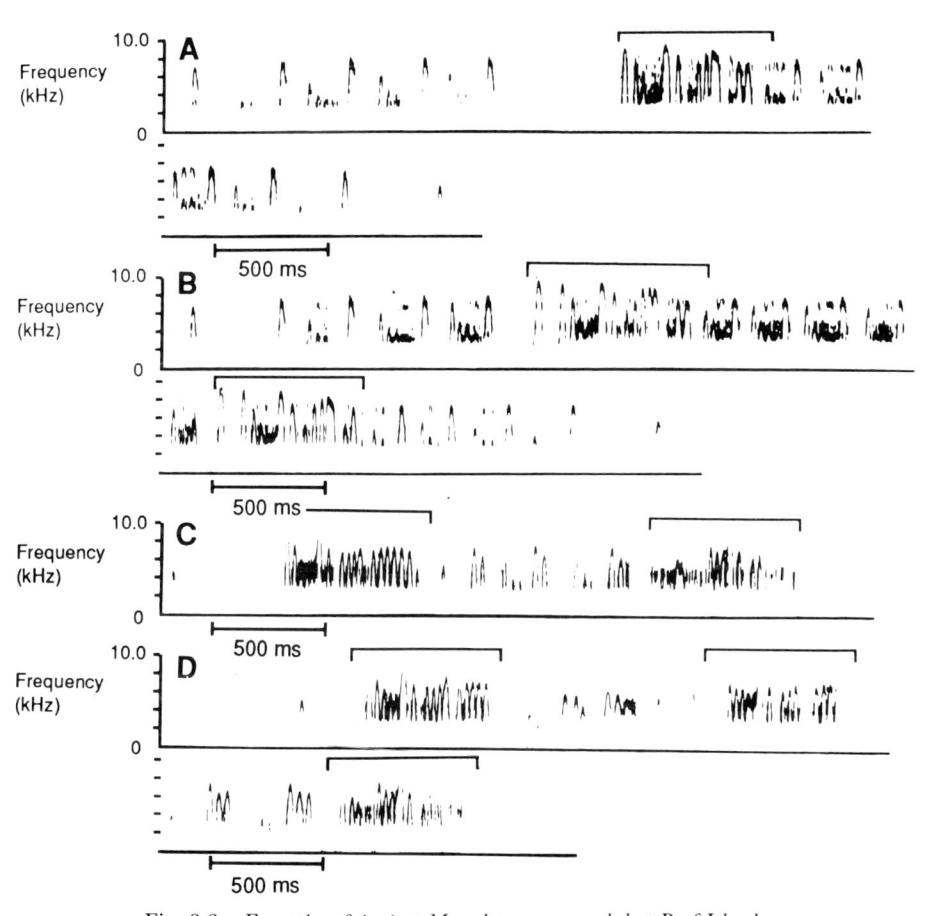

Fig. 8.3 *Examples of Ancient Murrelet songs recorded at Reef Island.*

songs. The songs generally lasted for several seconds, and were repeated at intervals of 20–30 s. Singing by one bird apparently stimulated others nearby, and bouts of countersinging could last up to 45 min.

None of the three other calls identified consisted of chevron-shaped elements. *Long-whistles* (Figure 8.4, A) and *short-whistles* (Figure 8.4, B) both consisted of pure tones, differing in length, while the *wheeze* was a short, unstructured call which came out on the spectrogram as a broad smudge (Figure 8.4, D).

PRE-LAYING

Seabirds must visit their breeding colonies to undertake a variety of activities associated with reproduction. Judging from the evidence of numerous pre-

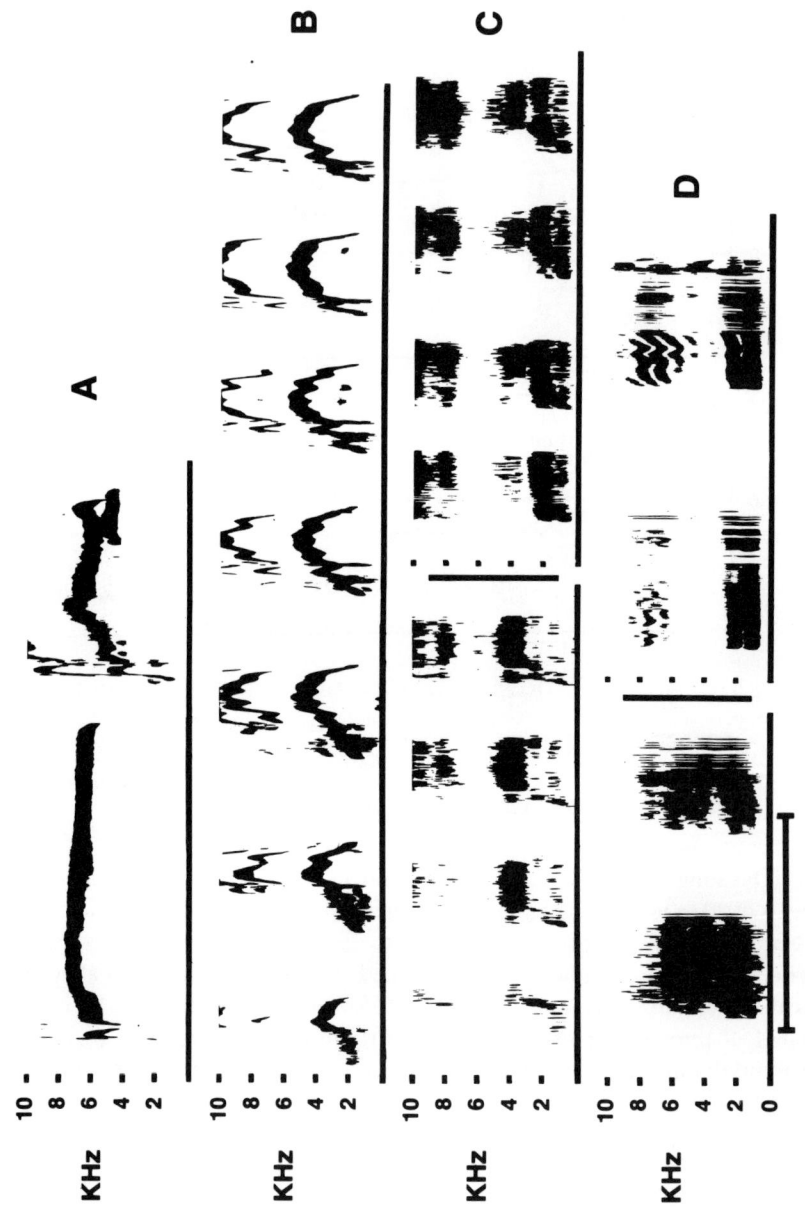

Fig. 8.4 Examples of other forms of Ancient Murrelet vocalizations recorded at Reef Island (A, long whistles; B, short whistles; C, D, wheezes and other atonal calls). Scale bar = 500 ms.

dations liberally scattered over most murrelet colonies, visits to land are fraught with greater danger for Ancient Murrelets than for most other seabirds. Consequently, many aspects of their behaviour seem to relate to the need to minimize their chance of getting caught.

We used a variety of techniques to obtain an index of the number of Ancient Murrelets visiting Reef Island each night, and to measure the amount of activity taking place. During 1984 and 1985 Ian Jones and his assistants maintained a continuous watch every night for at least 2½ h, from the time of the first arrivals, and counted the number of birds landing, and the number singing and calling, in a fixed area in the centre of the colony (Plot C). Sometimes we continued the observations right through to dawn. From 1985 onwards we measured activity on the offshore gathering ground by counting for 10 minutes the number of birds flying through the field of a telescope pointed from our lookout close to camp towards the navigation light at the north end of Low Island, about 4 km away.

In all years we placed small plastic tags in burrow entrances in such a way that they would be displaced by birds entering the burrows, and checked and replaced them daily. This technique has been widely used in studies of burrow nesting seabirds. It does not allow active breeding burrows to be distinguished from those merely being prospected by non-breeders, but it does give a measure of the overall amount of activity on a given night. Although mice definitely use the burrows from time to time it seems that they usually avoided the tags, because at the end of the season, on nights when few birds visited the colony, we found very few tags displaced.

Like other auks, Ancient Murrelets make only periodic colony visits during the period between their first arrival in spring and the start of laying. On some nights, the forest is completely silent, whereas on others birds shower out of the sky and the chorus of songs and calls is almost deafening. Generally, nights of big arrivals were calm and moonless, or overcast (Jones et al., 1990). If the moon set during the course of the night then there was often a wave of arrivals following its disappearance. Likewise, a bright moonrise curtailed most of the activity (Figure 8.5). Weather conditions also affected attendance during the pre-laying period, with few birds arriving at the colony when winds were strong (Figure 8.6). On stormy nights there was little or no activity on the colony.

During the pre-laying period, birds began to gather on the gathering ground about 3 h before sunset. This means that some birds spent at least 5 h waiting offshore before they came to land, because the first arrivals on the colony slopes usually occurred about 2 h after sunset. The timing of the first arrivals in relation to sunset remained fairly consistent throughout the season (Figure 8.7).

It was very hard to tell what was going on at the colony during the pre-laying period. We tried using image-intensifying, night-vision 'scopes to watch behaviour, but these proved relatively ineffective within the forest, because the

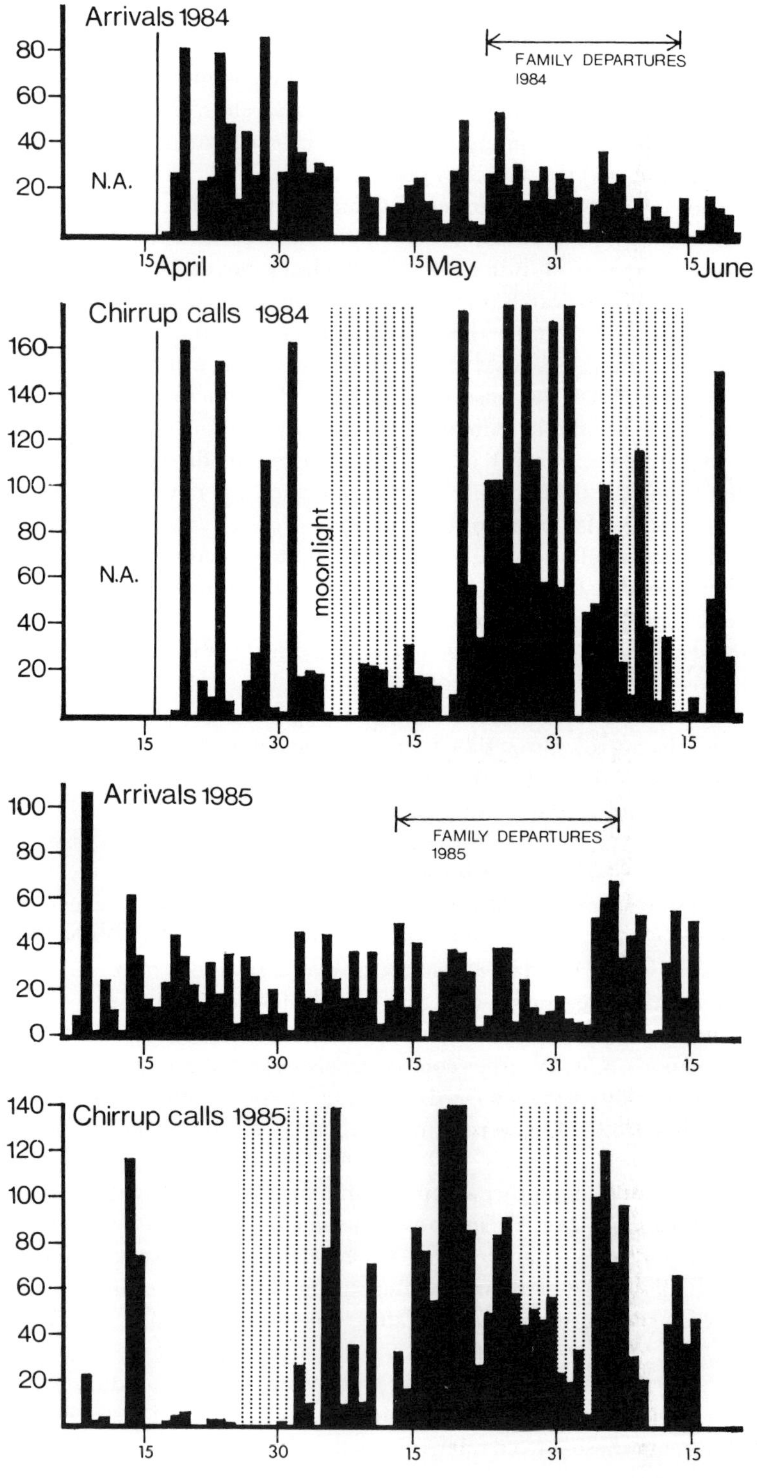
Arrivals 1984
FAMILY DEPARTURES
1984
N.A.
80
60
40
20
15 April
30
15 May
31
15 June
Chirrup calls 1984
moonlight
N.A.
160
140
120
100
80
60
40
20
15
30
15
31
15
Arrivals 1985
FAMILY DEPARTURES
1985
100
80
60
40
20
0
15
30
15
31
15
Chirrup calls 1985
140
120
100
80
60
40
20
15
30
15
31
15

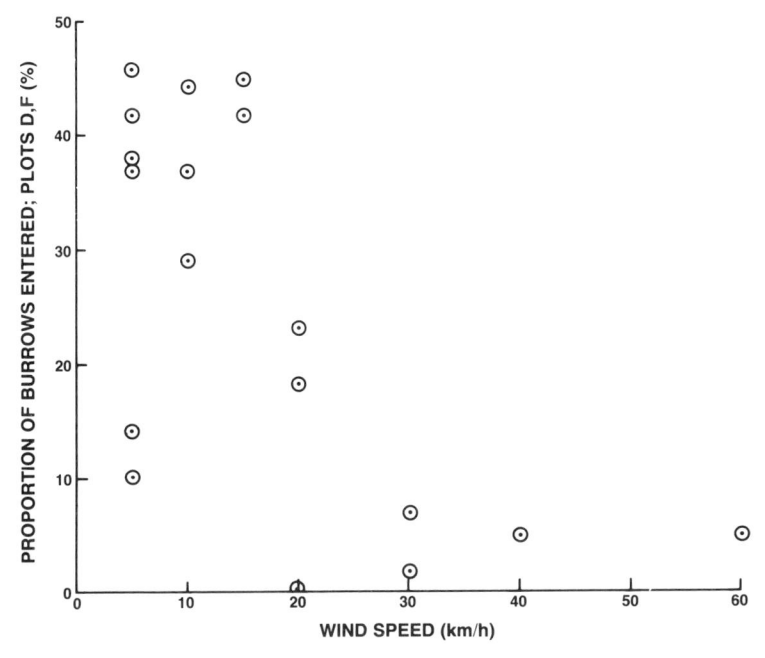

Fig. 8.6 *Numbers of burrows entered at two plots on Reef Island in 1989 in relation to the wind speed at midnight.*

available light was too low. Consequently, like owls, we relied heavily on what we could hear. If you sit listening in the forest on a calm night you first detect a few faint, far-off calls from murrelets flying above the tops of the trees. Then you may hear the fluttering of wings and the snapping of small twigs as arriving birds blunder through the canopy. Particularly if the night is not pitch black, many of the earliest arrivals perch above the ground, often in the lower branches of the canopy. After a few minutes of silence they then begin to call, using either the *chirrup*, or the *song*. We guess, from their vocalizations, that many of these early arrivals are males (see below), but there is little evidence on this. Some arrivals involved two birds flying in together and landing close to one another. These may have been pairs, because in some cases both birds entered the same burrow. After a short time, some of the arrivals begin to reach the ground. Whether the majority of those landing come directly from the sea, or whether most pause in the canopy before descending, is hard to tell, but probably a greater proportion of those arriving in the middle of the night fly straight in without pausing in the trees.

Within a half an hour after arrivals have begun, there are many birds on the ground and in the trees. Some sit on the surface calling, while others enter

Fig. 8.5 *Numbers of birds landing and chirrup calls recorded on Plot C at Reef Island in 1984 and 1985 during the first 2 h of arrivals.*

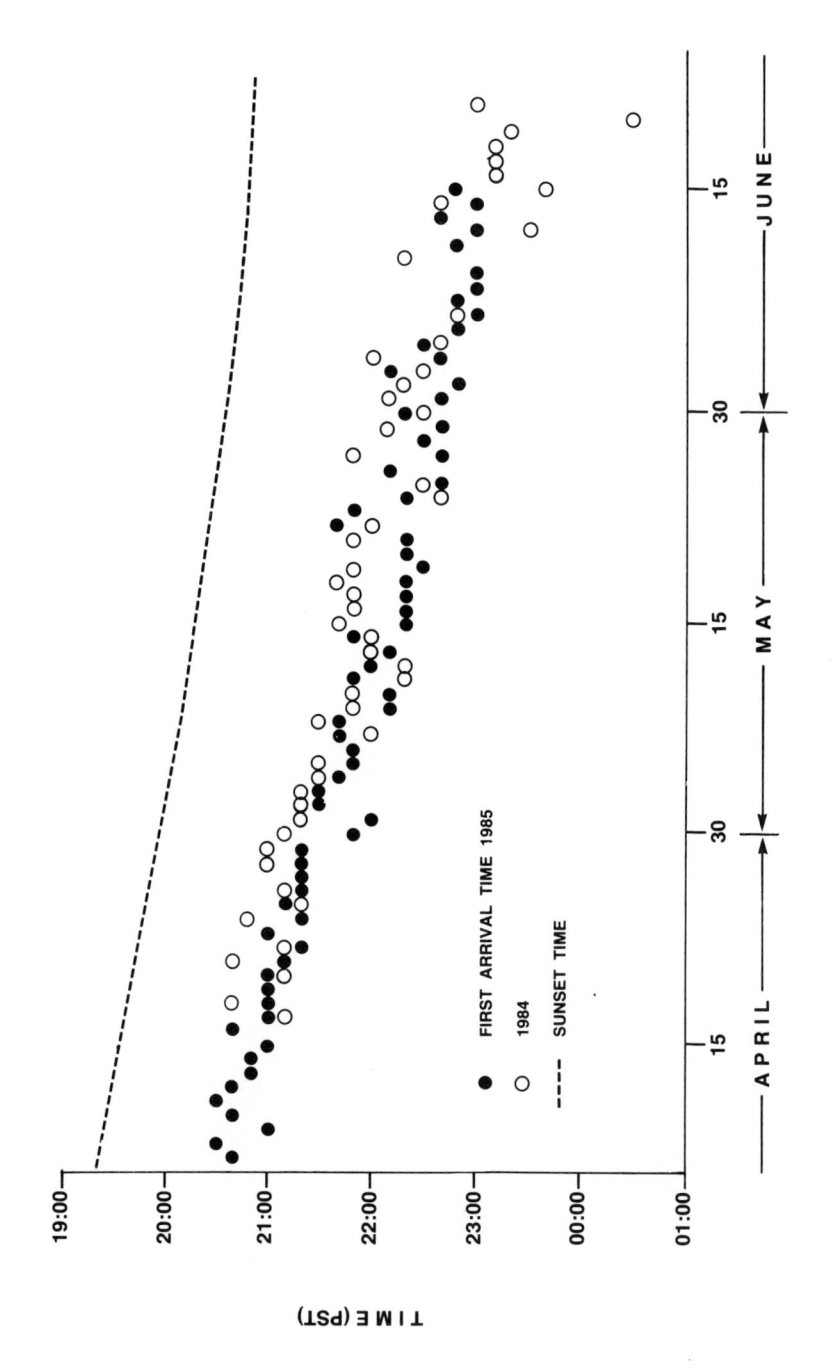

Fig. 8.7 The timing of first arrivals at Reef Island, in relation to the time of year (PST = Pacific Standard Time).

burrows, and from some of these you can hear a prolonged and rather sibilant series of whistling and clucking sounds, mixed up with *chirrups* and sometimes snatches of *song*. These prolonged vocal medleys we termed *chatter* calls. Typical songs are usually delivered only from more than 10 m up in a tree or, less often, from on top of a stump or fallen log. Singers may return to the same song-post for many nights; one bird with a distinctive song was recorded in the same spot on 25 nights during the 1985 season (Jones *et al.*, 1989). The following extracts from entries in the Reef Island log for 1989 give examples of typical nights;

> 28 March (light wind, no moon, but a faint aurora). "Gathering ground activities at 1800 hrs included many 'flop displays' [see below]. Most birds were in pairs. Many arrivals [to the gathering ground] were also in pairs or larger groups. Many birds arriving on the colony from 2140 onwards. At least a dozen arrived on plot D before 2200. Later, fairly quiet, with sporadic calling and a few birds singing. Lots of calling in burrows, including D4 [used in 1988, but not occupied in 1989]. It seems that early arrivals mainly went straight into burrows. Later, a few birds were found on the surface, though there had been few arrivals; these were probably birds that had emerged after having spent some time in burrows. Recent knock-downs [of tags at burrow entrances] have mainly been at burrows used last year."
>
> 31 March (wind moderate, cloud and light rain, no moon). "First bird caught at 2110; a female with egg. A few other arrivals at about the same time, then very quiet until further arrivals about 2130. Calling fairly brisk at 2300–midnight, tailing off afterwards. No rush or birds and very few seen on surface. Some burrow calling."
>
> 3 April (light wind and rain, no moon). "Very big night. First arrivals about 2130, then very busy to about 2330; afterwards quieter, with many singers in the trees and many calling from burrows, but few birds on the surface. Finish 0200."
>
> 5 April (wind moderate, no rain, no moon). "Another big night. Many birds arriving from 2145 onwards, much calling in the trees and also in burrows. Two birds seen interacting in a burrow entrance, giving the chatter call. Another bird was in a burrow entrance about 1 m away. The bird on the outside (i.e. closest of the couple to the burrow entrance) did not appear to be calling. When trapped, it had a bill-depth of 6.4 mm, clearly female."

During the pre-laying period, more than 75% of burrows in area F were entered at least once. On peak nights more than 50% were entered. Generally speaking, those where eggs were subsequently laid were entered more than others, but a few which were never used were entered almost nightly (Figure 8.8). As only 30–40% were eventually used, we assumed that a lot of prospecting was going on during the early part of the season. Presumably, previously mated birds were reuniting and new breeders, or those which had lost their mates, were trying to locate and suitably impress members of the opposite sex. However, the degree of activity on the surface during the pre-laying period was usually lower than was characteristic of the end of the season, when

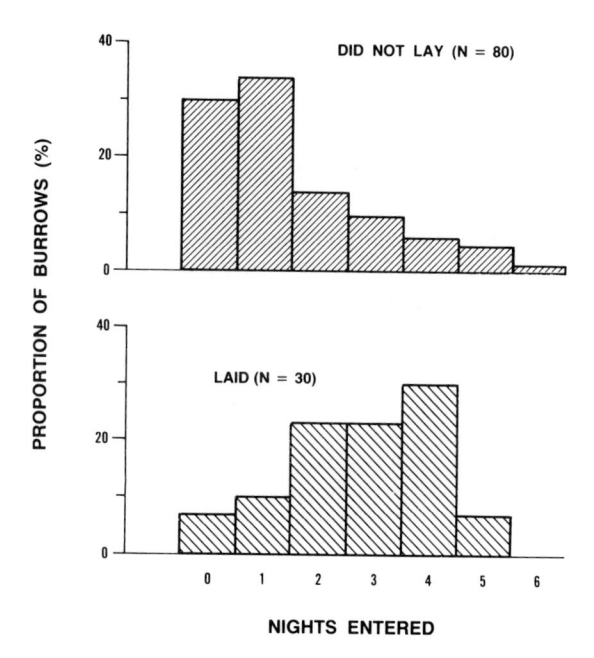

Fig. 8.8 *Number of nights in March 1989 that burrows at Plot F were entered in relation to whether or not they were used for breeding.*

many young non-breeders were visiting the colony. Also, there were few signs of new burrows being dug, although there was usually some evidence of burrows used in previous years being cleaned out, because old eggshell and egg membranes appeared at burrow entrances, and a few birds were trapped with muddy feet and claws. Probably many birds visiting the colony before egg-laying had begun were those which had bred in previous years and most of these birds must have used pre-existing burrows. As explained in Chapter 7, most birds trapped before the median date at which the second egg was laid, probably attempted to breed that year.

Fighting among Ancient Murrelets on the colony was observed occasionally. Like most auks, they grappled one another with their bills (Birkhead, 1985), giving shrill *trill-rattle* calls at the same time. We sometimes trapped birds with feathers missing around the base of the bill, and with cuts to the lower mandible, probably the result of fighting. Litvinenko and Shibaev (1987) observed some birds with blood on their breast feathers from wounds around the mouth, so the fighting can become very violent. We also trapped a bird with blood on its plumage, but the site of the wound was not apparent and the blood may have come from its opponent. Because there were many unoccupied burrows on Reef Island every year, it seems unlikely that competition for nest sites was very intense, and this may account for the relatively small amount of fighting observed.

A major mystery concerns where Ancient Murrelets on Reef Island copulated. They were very discreet about it, because we only once saw copulation on the ground, and we rarely saw birds sitting about on the surface in pairs. Litvinenko and Shibaev (1987) also saw one copulation, "on top of a rock", on the colony. Despite many hours of observations of the gathering ground we saw only one attempt at copulation there, between birds perched on a floating log (it was unsuccessful). Consequently, we assumed that copulations at Reef Island normally took place underground, in the burrow. Flint and Golovkin (1990) reported that, in the U.S.S.R., copulations were regularly seen on the surface of the colony, but they did not explain how the observations were made. It would be intriguing to know whether there is a real difference between Russian and Canadian murrelets in their candidness about sex, or whether we simply did not adopt the right approach.

EGG-LAYING

First eggs were usually laid at Reef Island in the first half of April, although a few were laid in late March in very early years. During the egg-laying period the females visiting the colony to lay eggs that night were usually the first to land (Figure 8.9). With practice, it was possible to tell from the sound of the

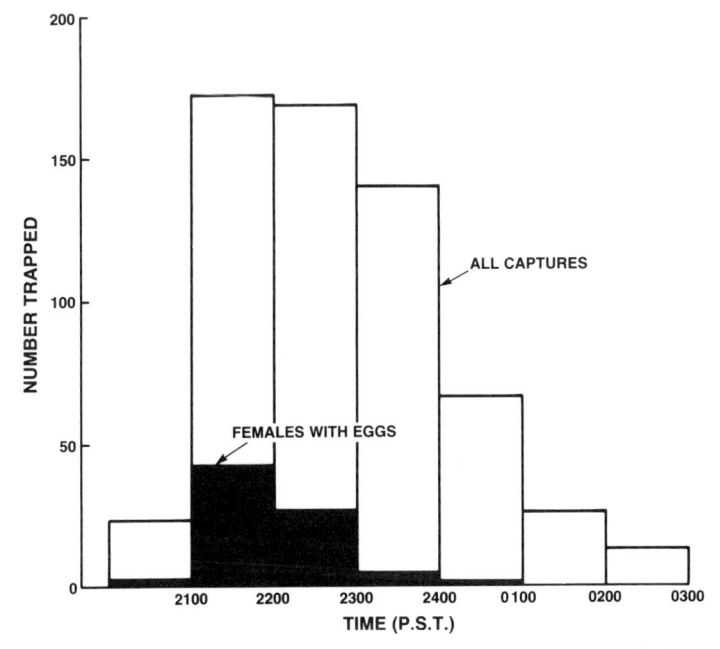

Fig. 8.9 *The number of egg-laying females trapped, in relation to the time at which other birds were captured (PST = Pacific Standard Time).*

116 Studies at Reef Island

landing whether or not the arriving bird was carrying an egg. Most breeding birds made a fairly soft landing, characterized by a whirr of wings and a subdued scuffling sound as they actually landed. Females carrying eggs generally hit the ground with a much louder thump. They were also much slower to take off, if approached with a light, and if the ground was uneven they obviously found it difficult to fly, such was the size of the egg with which they were burdened.

Sealy (1975b, 1976) considered that the breeding females did not visit the colony between the laying of the first and second egg, and there would seem to be little reason for them to do so, provided that they store the sperm for the fertilization of the second egg during the interim. However, we did catch one female at the colony between layings, indicating that at least a few do make visits during this period. The arrival of females ready to lay was unaffected by weather conditions, or the numbers of other birds visiting the colony (Figure 8.10). This suggests that laying cannot be delayed in response to adverse

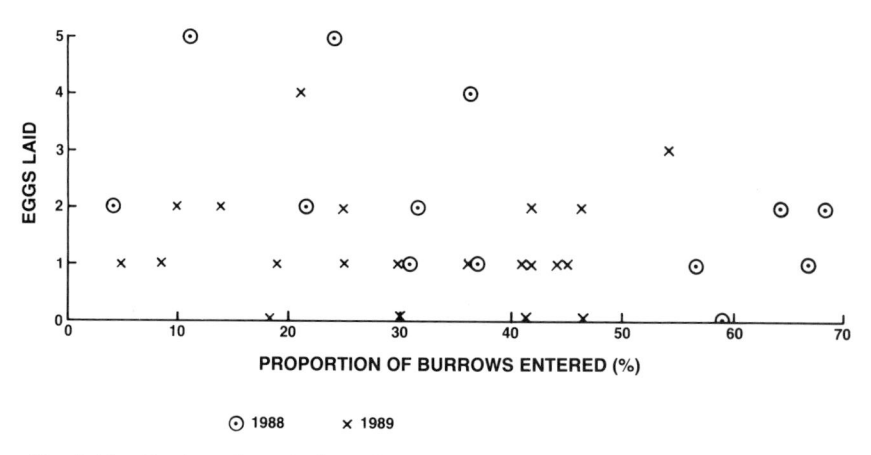

Fig. 8.10 *Numbers of eggs laid, in relation to the proportion of burrows entered that night.*

environmental conditions. Hence, females coming in to lay may have to contend with high winds shaking the canopy and making it hard for them to navigate to their burrows. Every year we found a few eggs on the surface, and some of these may have been laid by females that were unable to locate their burrows. The need to have some light to locate their burrow may have a bearing on the fact that females with eggs tend to be the earliest to arrive on the colony each night. Tags at burrow entrances were displaced nearly every night between the first and second layings, so it appears that males make frequent visits at this time (Figure 8.11).

Singing and surface activity declined as the number of pairs incubating rose, so that by late April in most years there was a period of relatively little activity

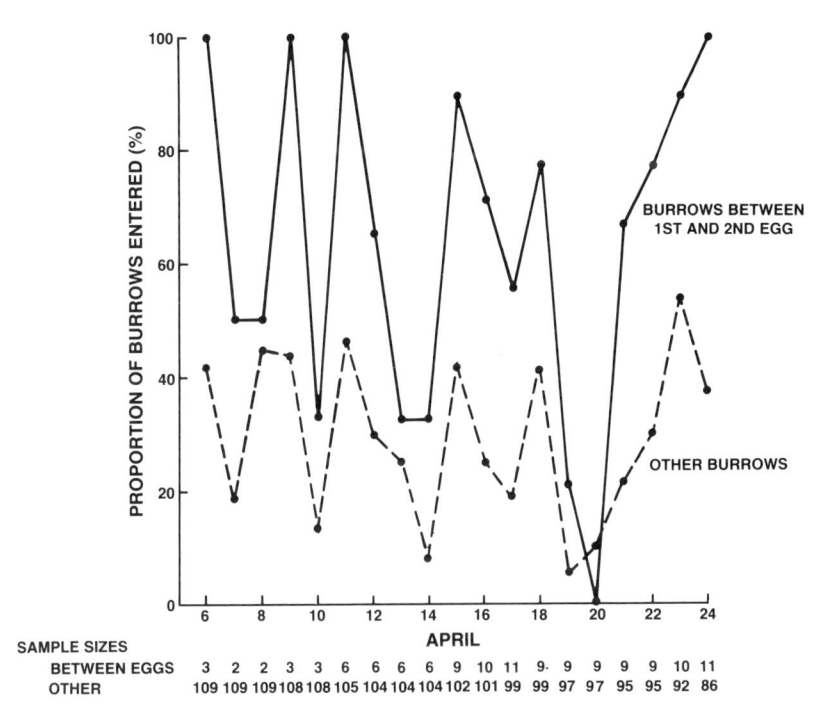

Fig. 8.11 *Visits to burrows between the laying of the first and second egg in 1989, in comparison with visits to burrows where eggs had not been laid.*

at the colony. This lasted until about the middle of May, when non-breeding birds began to arrive in big numbers (Chapter 9).

INCUBATION

By the beginning of May, most breeding pairs at Reef Island were incubating. In 1984 and 1985 we investigated the duration of incubation shifts by marking one member of each of several pairs with a spot of paint on the wings or back. Change-overs occurred at intervals of 1–6 days (Figure 8.12). In some pairs, such as D4 in 1985, one member of the pair undertook much more incubation than the other. At F18 in 1985, one bird apparently completed the final two weeks of incubation unaided, perhaps because its mate deserted, or was taken by a predator. However, at most burrows, the pair shared incubation fairly equally, so it seems unlikely that one sex habitually incubates much more than the other. There seemed to be a slight tendency for shifts to become shorter as incubation progressed, but the trend was not statistically significant. Once the chicks had hatched, one-day shifts often occurred. Possibly the

118 Studies at Reef Island

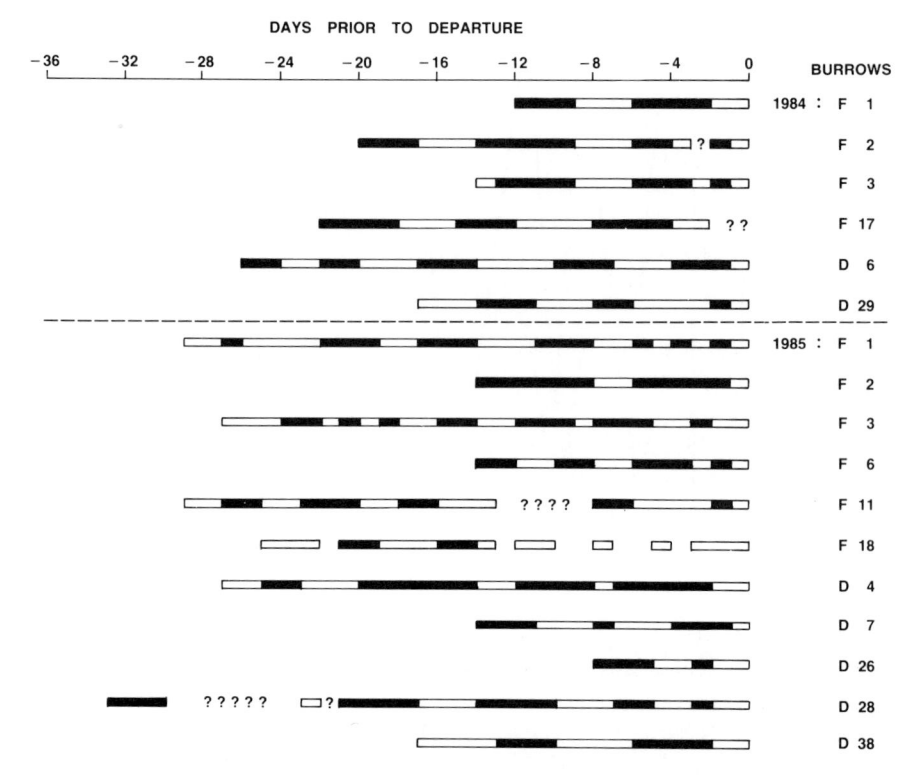

Fig. 8.12 Duration of incubation shifts at burrows where pairs were individually marked.

off-duty parent visits nightly at this stage, to be sure of being present when the chicks depart (see below).

In 1984 the mean duration of 38 complete shifts was 3.1 days, and 82% lasted 3 days or more. In 1985, the mean duration of 122 shifts was 2.3 days, and only 38% lasted more than 2 days (Table 8.1). Sealy (1976) found that the lengths of incubation shifts at Langara Island in 1971 were very similar to

Table 8.1 *Lengths of incubation shifts at Reef Island, and the proportion of incubation time spent in shifts of different lengths. A median test comparing the lengths of shifts in 1984 and 1985 (≤ 2 days vs >2 days) showed that shifts were significantly longer in 1984 ($\chi^2 = 22.3$, $P < 0.001$).*

Year	1	2	3	4	5	6	n	Mean
1984	1 (1%)	6 (10%)	19 (48%)	11 (37%)	1 (4%)		38	3.13
1985	26 (9%)	50 (35%)	34 (36%)	8 (11%)	2 (4%)	2 (4%)	122	2.31
Totals	27	56	53	19	3	2	160	

Length of shift (days)

those that we observed at Reef Island. He reported that all lasted three days. However, his Figure 7 shows that some were shorter than this, and one was of four days. He apparently confined his analysis to shifts which began and ended with a change-over, excluding those following or preceding a period of neglect. As at Reef Island, he found that the frequency of change-overs increased at the time of hatching.

We obtained little information on whether off-duty breeders visit the colony without exchanging incubation duty. At some burrows we found tags knocked down without any change-over taking place. We also observed that incubating birds sometimes came to the burrow entrance at night, and this may have caused such tag displacements. Bendire (1895) mentioned that the incubating adult sometimes emerged in the evening to call. Two radio-tagged breeders did not visit the colony except to exchange incubation duty (total of eight visits), but one did return to the island on one occasion after being relieved and flying down to the sea (probably very close to the shore). On returning, after only 20 minutes on the water, it landed very close to its burrow, and definitely went underground (the signal becomes very weak when the bird enters a burrow). However, we don't know whether it entered its own, or another, nearby burrow. It left again after 20 minutes. Its behaviour indicates that not all breeding birds arriving at the colony, particularly late in the night, are necessarily returning from one or more days of feeding. Neither of the radio-tagged breeders was detected on the sea around the colony until after dark. The transmitters could be detected up to 6 km away, so this indicates that they did not spend much, if any, time on the gathering ground. However, the sample was very small, and the birds were burdened with the transmitters, so their behaviour may not have been entirely typical. It is possible that some breeders do spend substantial periods on the gathering ground, before arriving at the colony.

Breeding birds arriving at the colony to take over incubation duty may reach their burrows at any time during the night. On first entering, there is an immediate burst of chatter calling by both members of the pair. We recorded these calls by inserting into occupied burrows microphones attached to a tape recorder recording at slow speed. Replaying the tape allowed us to tell when the bird entered, and how long the initial calling lasted. The longest continuous bout of greetings lasted 30 minutes, delivered with only the briefest pauses, and most greetings lasted several minutes. We presumed that this exchange of calling functioned to maintain the pair bond. Similar calling is heard from non-breeders during courtship (see below).

DEPARTURE OF FAMILY PARTIES

Normally both parents are present when the chicks leave the burrow. As there is never more than one breeder present in the burrow during the day, the departure necessarily takes places only after the arrival of the off-duty bird (Litvinenko and Shibaev, 1987). The timing of departures of all family parties leaving from plot C in 1984 and 1985 was recorded by Ian and his assistants (Jones *et al.*, 1987b). The peak came within the first hour after the earliest birds began to arrive at the colony, and more than 90% of departures took place within the first 2 h. The arrival of chicks at our trapping funnels gave a good idea of the timing of departures, and how it changed over the season. In mid-May, with complete darkness by about 22.30 h our first chicks were reaching the beach between 23.00 and 23.30 h, with the peak numbers coming at about midnight. By the end of the month, with darkness coming half an hour later, the average first arrival was about midnight, with a peak close to 01.00 h (Figure 8.13). Like other aspects of behaviour, the timing of departures was much affected by light conditions, with chicks reaching the beach earlier on rainy, overcast nights than on clear nights. This effect was most obvious between 17 May and 7 June, when an adequate sample of chicks was trapped

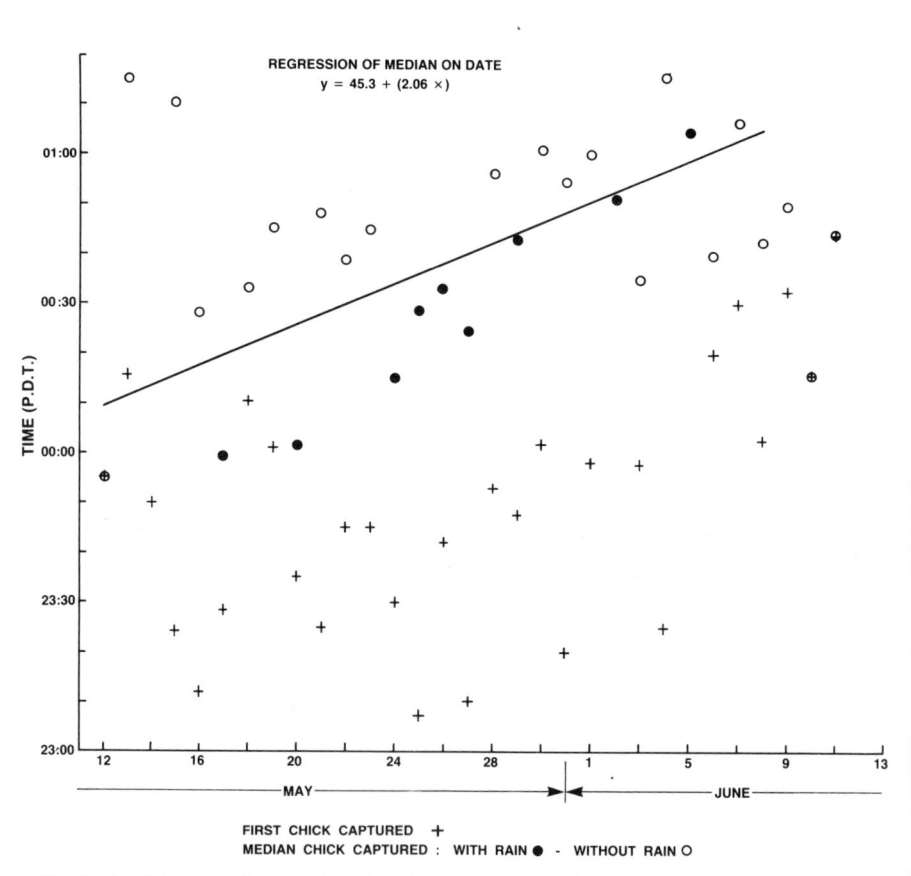

Fig. 8.13 *Timing of first captures of chicks at the trapping funnels, in relation to time of year.*

every night. During this period the median time of chick capture averaged 18 minutes earlier on nights with rain than on nights without. This difference presumably is determined by the timing of arrival of the incoming adult, which waits for complete darkness before landing.

We were able to determine the course of events during departure by listening carefully on still nights, without moving or using a light. The following description is taken mainly from Jones et al. (1987b). Departures were preceded by *chatter* and *chirrup* calling inside the burrow for a maximum of 21 minutes (average of 19 departures, 4 min), after which one or both parents left the burrow and called from just outside the entrance. The vocalization given at this moment was a very emphatic version of the chirrup call. With practice, it was easily recognizable. It was repeated at intervals of about 12 s. After one or two such calls the chicks responded with a thin, but urgent, "pee-pee-pee", or "pee-pee-pee-pee", in tone not unlike the disyllabic contact call given by young murres at the time of their departure from the breeding colony (Tschanz, 1968). This exchange of calls, by all members of the family, continued for up to 10 minutes (average of 33 departures, 4 min), during which time the whole party emerged from the burrow and moved together up to 20 m (usually less than 10 m) towards the beach. The parents then flew off to the sea, sometimes pausing briefly in a tree to give a few more calls on the way. Once on the sea they sat on the water 50–200 m from shore and continued to give the same emphatic *chirrup* given at departure from the burrow. Birds observed through the night 'scope swam actively backwards and forwards while calling. Every few minutes they took off and flew 50–100 m before landing again and continuing to call. We had the impression that some birds patrolled as much as 400 m along the shore. This may have been especially necessary near our camp, because the lie of the ground could have channelled chicks from certain areas in either of two directions, bringing them out on either side of the camp promontory.

Once their parents had departed, the chicks ceased calling within one or two minutes and began to run rapidly towards the sea, scrambling over logs and fallen branches with great agility. They made no attempt to remain together, and without calling this would certainly have been impossible in the darkness. We sometimes trapped the two members of brood that had been banded in the burrow in different catching funnels, showing that their seaward routes could diverge considerably. Shibaev (1978) gave a nice description of the chick's activities once on its own:

> The young bird crouches as it runs, balancing itself with its wings. Its actions as it moves . . . are energetic and purposeful. It jumps off rocky ledges, ascends with springs, and as it scales large mounds of rocks it squeezes through narrow clefts and tirelessly searches for alternate paths if an obstacle is insurmountable.

We trapped and banded chicks just after the departure of their parents in areas 150–250 m from the sea, and observed the time that elapsed before they reached the trapping funnels. It took them from 12 to 30 minutes, the two members of one brood being separated by as much as 14 minutes, although about half (8/18 broods) arrived within one minute of one another. During this period the parents were presumably sitting on the sea calling.

On reaching the sea, the chicks plunged in immediately. They frequently made one, or more, brief dives, before paddling furiously away from the shore in search of their parents, often in a series of zig-zags. However, a flashlight pointed from the shore was likely to divert them from their course, and some chicks walked back up the beach in response to a strong light.

In 1987 we radio-tagged one member of each of five breeding pairs, after their chicks had hatched. This allowed us to follow the whole sequence of events from a distance. Two of the birds involved exchanged incubation duty with their mates on the night after they were tagged, and both returned the following night, when the departure took place. Hence they had been absent from the colony for less than 24 h when they returned. Contact with both was lost by 06.30 on the morning following their incubation exchanges and on the next night (the night of the family departure) neither bird was detectable until just before they landed on the ground. Presumably they had not spent much time on the gathering ground after leaving the colony, or before arriving the next evening. The radio signals indicated that they entered their burrows within one minute of landing, and both emerged again within 10 minutes. All five radio tagged birds spent less than 5 minutes between leaving the burrow and flying down to the sea and, although it was difficult to pinpoint them exactly, none appeared to move more than a few metres from the burrow before flying off. Once on the water, they all remained close to shore, moving backwards and forwards along about 200 m of shoreline, although we lost contact with one bird entirely for about 5 minutes, probably indicating that it had flown to the far side of the island.

The chicks from two burrows within the catching area of our funnels were trapped on their way to the shore about 20 minutes after their radio-tagged parents reached the sea. After about another half-hour the radio signals began to get slowly weaker, indicating that the families had assembled and were moving away from the island. Contact was lost, in a northeasterly direction, after a further 2 h (Duncan and Gaston, 1990).

The beach and the surf appear to be considerable obstacles for the chicks, as Heath (1915) observed at Forrester Island. At Reef Island the boulders are anything from 20 to 100 cm across, and the chicks floundered among them for a long time when the tide was low. Once in the sea, the chicks generally swam strongly, churning along with their huge feet at amazing speed—"like a motor boat", as Shibaev (1978) put it. They were capable of diving immediately. When we trained spotlights on them, while making a film of their departure, they could be clearly seen swimming underwater, using their wings as paddles, just like the adults. On nights of rough seas, some chicks were repeatedly cast back on shore by the surf. Each time they would pick themselves up and rush back into the water to try again. Willett (1920) gave a graphic account of this behaviour:

> At midnight, with the aid of a lantern, the writer has watched these downy chicks [Ancient Murrelets], not more than three or four days old, dive through the surf in response to the cry of the parent bird . . . and this at a time when boulders weighing a hundred pounds or more were being rolled up and down the beach like so many pebbles.

The beaches where we watched Ancient Murrelets departing were relatively sheltered. On Reef Island most of the worst gales come from the southeast, so the main colony, along the north coast, gets much less surf under most conditions than the south coast. Despite this, we saw chicks take a terrible pounding in the most severe weather, although none actually appeared to be injured by being rolled in the surf. On the west coast of the Queen Charlotte Islands, subject to big swells from the Pacific Ocean most of the time, the main colony areas tend to avoid the exposed, west-facing coasts of the breeding islands (M. Lemon, pers. comm.). The amount of protection afforded by the adjacent departure beaches may be a factor in the Ancient Murrelet's choice of breeding sites. However, the forest also tends to be taller along sheltered shores, and this could also influence their choice of colony area.

Apart from the surf, the chicks have a variety of other dangers to contend with before they reunite with their parents. Saw-whet Owls were present on Reef Island every year, and we found the down of Ancient Murrelets in their pellets. Deer Mice killed several chicks when they were left unattended in burrows, and they also attacked them on the way to the sea, biting them on the neck and back. In 1985 we saw several chicks attacked by mice as they arrived at our catching station. The attacks were drawn to our attention by a piercing scream of panic, uttered by the chick; we never heard this call otherwise and they never gave it when we were handling them. One chick "froze" when attacked. It seemed to be dead when we picked it up, but two minutes later it came back to life abruptly and began to behave normally. There was no sign of a wound on it. Because our funnels were concentrating many chicks in a small area we felt that the mice could become a problem. We baited some mouse traps with peanut butter, and after a few mice had been removed we saw no further predation. It is possible that our observations all involved a single, predatory mouse. We do not know how often chicks are ambushed by mice under normal conditions.

Once the chicks arrive at the sea they are not clear of danger. We saw two incidents through the night scope which we interpreted as fish attacks; in both cases a chick swimming away from land on a calm sea disappeared in a sudden swirl of water, and was not seen again. Howell (quoted in Bent, 1919) also saw Ancient Murrelet chicks attacked by fish while leaving the colony. Ian Jones saw a Harbour Seal surface underneath a chick and flick it into the air with its nose. The chick was apparently unharmed, but other seals might have been less kind.

HOW THE CHICKS FIND THEIR WAY TO THE SEA

In 1986 we carried out a series of simple experiments to discover how the chicks found their way to the sea. We had noticed that they were strongly attracted to light and we thought that they might also use the slope of the ground and the sound of the sea, and of adults calling, to find their way. As the chicks arrived at the catching funnels we placed them, one at a time, in a T-maze (Figure 8.14). This gave them the alternative of heading either towards, or away from, a test stimulus placed at one end of the 'T', which consisted of (a) very faint light from a masked flashlight, (b) a slope of 5°, or (c) 10°, (d) a recording of surf made in daytime, without birds, or (e) a similar recording made at night with several parents calling. The tests were carried out in a light-tight hut. Chicks were placed in the maze immediately after we

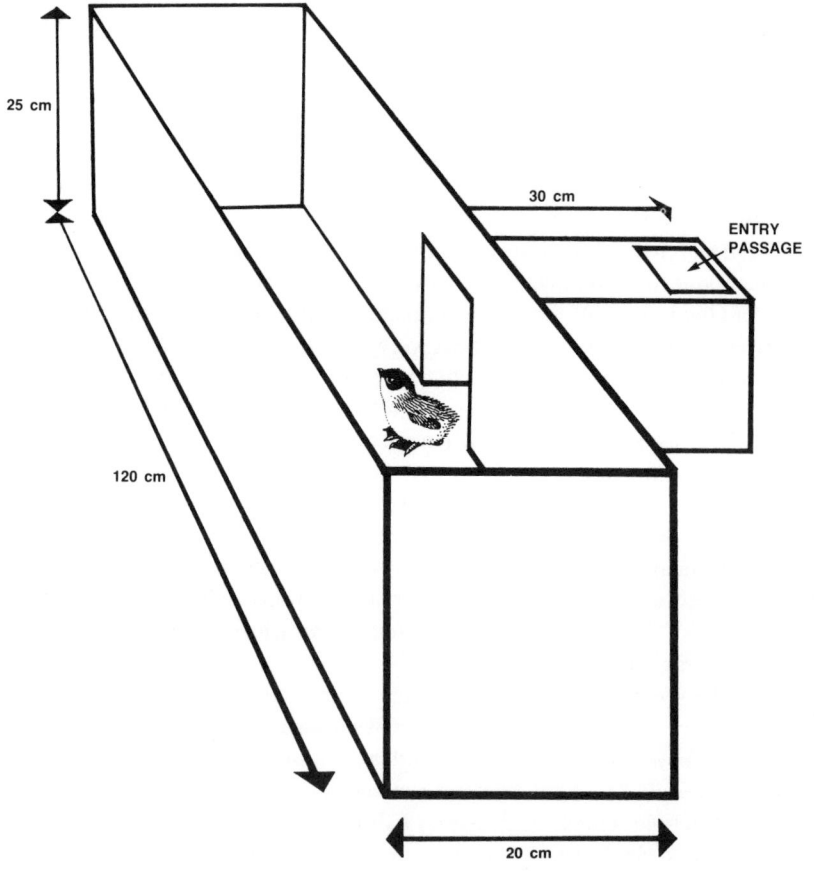

Fig. 8.14 *Apparatus used to test the ability of chicks to use different cues in finding their way to the sea.*

captured them in our funnel traps, and we tested each chick only once. We recorded the direction in which it headed, whether or not it reached the end of the maze, and how long it took to do so. All tests were stopped after 30 s and the chicks released to the sea (Gaston et al., 1988b).

As we expected, light proved to be a potent stimulus, with all the chicks choosing to head towards it, rather than away from it, most of them rather quickly. Responses to other stimuli were less constant. We found a significant tendency for chicks to orient towards the sound of the surf, but it made no difference whether or not adults could be heard calling. A slope of 5° made no difference to their choice of direction, but when it was 10° they showed a significant tendency to head downhill. We concluded that the most important cue used by the chicks to find their way to the sea was light intensity, which was invariably brightest over the water, but that they also responded to the sound of the sea and the direction of the slope.

Most previous accounts of the departure of Ancient Murrelet families have described the parents leading the young to the sea (Heath, 1915; Sealy, 1976; Shibaev, 1978; Litvinenko and Shibaev, 1987), but we did not find this to be typical at Reef Island. Some family parties were regularly found close to the shore, with parents and chicks calling to one another, but these may have originated from burrows close to the sea. They formed only a small minority of the chicks reaching the shore. Family parties moving together were sometimes encountered on the slopes, but it was impossible to tell what proportion these constituted, because once the parents left, the movements of the chicks were almost impossible to detect.

The description of Willett (1915), quoted in the introduction, suggests that events at Forrester Island were generally similar to those seen at Reef Island, with the chicks travelling to the sea alone. However, Mike Rodway and Moira Lemon (pers. comm.) who observed many departures at Rankine Island, also in the Queen Charlotte Islands, found many family parties arriving at the shore together. Rankine Island is very flat, compared with Reef Island. This may make the orientation cues used at Reef Island less effective and encourage the parents at Rankine Island to accompany their chicks for a greater proportion of the journey to the sea. This seems to be confirmed by Shibaev's account of departures from Verkovskii Island, in Peter the Great Bay. There, the parents remained with the chicks if the route was level, but flew to the sea if there was difficult terrain to be negotiated.

Sealy (1976) found some chicks arriving at the shore before any adult birds had landed on the colony that night. We also had several cases of chicks arriving at our trapping funnels before any adults had been heard calling that night, either on the sea, or on the colony. These may have been chicks that failed to reach the sea the previous night and remained hidden during the day. Michael Rodway and Vernon Byrd (pers. comm.) both reported finding chicks hidden under logs or vegetation, clearly not breeding sites, during the day, but we never saw this on Reef Island.

Chicks may sometimes initiate departure in the absence of either parent. We had no evidence of this for Ancient Murrelets, but Zoe Eppley, a researcher at University of California, who raised chicks of the related Xantus' Murrelet in captivity, made an interesting observation on the subject. Her chicks were hatched in an incubator. After hatching, she kept them in shoe-boxes, suitably lined, in her apartment. For the first two days they were docile, remaining quietly in their shoe-box burrows. However, about 48 h after hatching, suddenly they all scrambled out of their boxes and proceeded to rush frantically to the furthest parts of the house, turning up in every unlikely place that they could crawl into. A similar observation was made by Vermeer and Lemon (1986), who found that Ancient Murrelet chicks hatched from eggs placed beneath Cassin's Auklets, left their burrows after an average of 3.5 days, without apparently making any attempt to eat the food offered by their foster parents. Such behaviour strongly suggests that the chicks are innately programmed to depart after a certain period, irrespective of the presence of their parents.

Litvinenko and Shibaev (1987) mentioned that Ancient Murrelet chicks become much more active in the last few hours before they leave the burrow. We noticed that chicks removed for banding within 24 h of hatching were generally docile and remained immobile when placed on the ground, in strong contrast to their frantic struggles when captured during departure. After trapping at the funnels, we found that even chicks kept in a cloth bag in a dark place never ceased jumping and wriggling. Once departure has begun, they appear to enter a state of hyperactivity terminated only when they are reunited with their parents.

. . . AND HOW THEY FIND THEIR PARENTS AGAIN

Because most parents do not accompany their chicks to the sea, the members of the family have to have some way of recognizing one another to ensure a reunion. At the peak of departures there are many chicks leaving the shore at about the same time and many parents calling at sea. Observations with the night 'scope showed that chicks heading out to sea would swim straight past some calling adults, and make directly towards particular individuals, which we presumed to be their parents. At the same time, an adult often showed signs of recognizing their chick, by swimming rapidly towards it. An excerpt from my field notes for 29 May 1985 gives an example of a typical departure from land, observed through the night 'scope:

> 00.50 h. Two chicks released. Ran into dense floating weed [kelp] on the surface within 10 m of shore. One passed close by a calling adult and both dived briefly, but the chick carried on. Took about 1 minute to clear 10 m of weed, then travelled rapidly out. 80 m from shore one chick was "bombed" twice by an adult which dived beneath it and came up under it, whereupon the chick dived briefly.

A brood of day-old Ancient Murrelet chicks removed from a burrow at Reef Island for ringing.

At 100–120 m from shore the chick joined a different adult, then within 10 m the second chick caught up and within another 5 m another adult joined the group. All four then swam off keeping closely together. Lost to view about 250 m out.

On some nights, observations through the night 'scope suggested that more than half the birds in the zone frequented by parents awaiting their chicks were not involved in family departures. While the putative parents were generally solitary, and called emphatically at regular intervals, the non-parents were usually in small flocks of up to eight birds, called less often and socialized with one another. These small flocks sometimes flew off together.

A peculiar feature of the chicks' progress, while searching for their parents at sea, was the harassment that they received from adult birds (as in the above quotation). On several occasions adults were observed to dive beneath chicks and surface directly below them. Sometimes a chick was attacked in this way by several adults in succession, or even simultaneously. No chick sustained injury as a result of this treatment, and they usually swam on immediately. The adults involved were not giving the typical parental recognition call, and hence were presumably not in search of chicks of their own. This harassment of departing chicks is also seen during the departure of murre chicks from their colony. Those which lose contact with their parent are often surrounded by adults which peck at them and, like the Ancient Murrelets, torpedo them from below (Harris and Birkhead, 1985; pers. obs.). I have seen young murres chased and pecked to the point of exhaustion. No satisfactory explanation has been advanced for this behaviour in either species.

In 1985 Ian Jones conducted a series of experiments to test whether the chicks and their parents were capable of recognizing one another by call. Close to the shore, he built a circular tank, about 3 m in diameter, out of plastic sheeting. It was flooded with water to a depth of 10 cm and loudspeakers were fixed at opposite sides. To test whether the chicks could identify their parents, he made recordings of the parents calling as the family was leaving the burrow, then captured the chicks and took them down to the tank. The parental calls were played through one speaker, alternating with those of a strange adult, also made at burrow departure, coming from the other speaker. The effect was extremely convincing. Chicks paid no attention to the calls of the unrelated adults, but rushed immediately to the speaker from which their own parent's call was coming, and jumped up and down beside it, as though trying to get inside the speaker (Jones et al., 1987a).

To test the parents, Ian marked them as the family was departing from the burrow, by creeping up to them and gently pressing a piece of sticky, reflective tape on their backs. He then captured the chicks, recorded their calls and played them back through a loudspeaker placed close to the water at the tip of a rocky promontory. This test could be carried out only twice. On the first occasion the bird with the reflective tape approached the loudspeaker close enough to be recognized in the beam of a flashlight. The second time an adult called persistently just out of visual range. Ian recorded the call and, by comparing sonograms, was able to identify the caller as one of the parents. On both occasions there were many parents calling for chicks on the water, but only the two parents involved approached the loudspeaker.

Chicks seldom fail to reunite with their parents, if they survive to reach the waters beyond the breakers. We had only five records in six years of lone chicks seen at sea after dawn. However, in bad weather, some chicks may be driven on shore, or become exhausted before dawn, in which case we would probably not have seen them. In three instances the lone chicks continued to swim back and forth parallel to the shore and 100–200 m from it for more than an hour. Moira Lemon (pers. comm.) has seen similar behaviour among chicks seen alone in daylight. These observations suggest that the initial rapid departure from the beach does not continue for long. At some point their innate tendency to swim away from land must switch off, otherwise chicks that failed to contact their parents as they swam out would simply continue towards the horizon. Apparently it is replaced by a searching behaviour that maintains the chick at a more or less fixed distance from land, in the zone where it is most likely to encounter its parents.

THE FIRST DAYS AT SEA

Once the family party has reunited on the sea they move away from land very rapidly and it is rare to see an Ancient Murrelet chick within sight from the coast once it becomes light. In 1987 we placed small radio transmitters on chicks (0.8 g) and parents (2.6 g), gluing them to the feathers, or down, of the back. We tracked them with a directional antenna, both from Reef Island, and from a light plane, in which David Duncan doggedly quartered the waters of Hecate Strait (Duncan and Gaston, 1990).

David managed to relocate 11 birds within 24 h of their departure from Reef Island. The seven found 6–8 h after leaving (07.00–09.00 h) had travelled an average of 13 km, at an average speed of just under 2 km/h. Surprisingly, the average speed of those five located later in the day, 12–18 h after departure, was considerably higher, averaging 3 km/h. The lower average speed of the early stage may be due to the inclusion in the calculation of the time spent by parents and chicks in finding one another.

The main surface currents in Hecate Strait are tidal in origin, and hence would have reversed over 12 h. Consequently, it seems that the speed of almost 1 m/s attained by family groups resulted mainly from their own locomotion. The maximum speed achieved by a surface vessel is constrained by its waterline length, because of the turbulence created by its bow wave. A similar constraint applies to a bird swimming on the surface, and Prange and Schmidt-Nielsen (1970) showed that, for a duck, which had a waterline length almost double that of an Ancient Murrelet, the energy expended on swimming increased very steeply at speeds above 1 m/s. It seems that adult Ancient Murrelets were probably swimming flat out during the period immediately after leaving the colony, and the speed of the chicks can only be accounted for by suspending the laws of physics. However, the chicks ride very high out of the water, and almost certainly hydroplane over the surface when going at full speed. This enables them to circumvent the limit imposed by their waterline length of about 6 cm.

David found only four family parties more than 24 h after departure, and these were at about the same distance as those located 12–18 h after departure, indicating that the phase of rapid dispersal may cease after about 18 h. The positions of those located suggested that the initial dispersal from Reef Island was mainly towards the northeast, but that once the initial period of rapid movement was over, the broods dispersed themselves widely in central Hecate Strait (Figure 8.15).

Although we banded 310 breeding adults with chicks during the course of the study, only one of these was retrapped again on the colony in the same season. None of our radio-tagged birds reappeared at the colony within the 2–3 week life of their transmitter batteries. In early June, after most chicks had left, we caught several birds with full-sized brood patches that were beginning to refeather. One was singing in a burrow, while two others showed signs of having been engaged in burrow excavation. These birds might have been failed breeders, or might have returned to the colony after losing their chicks at sea. By mid-June a few could have reared their chicks to independence, because Sealy (1976) saw independent young-of-the-year near Langara in

130 Studies at Reef Island

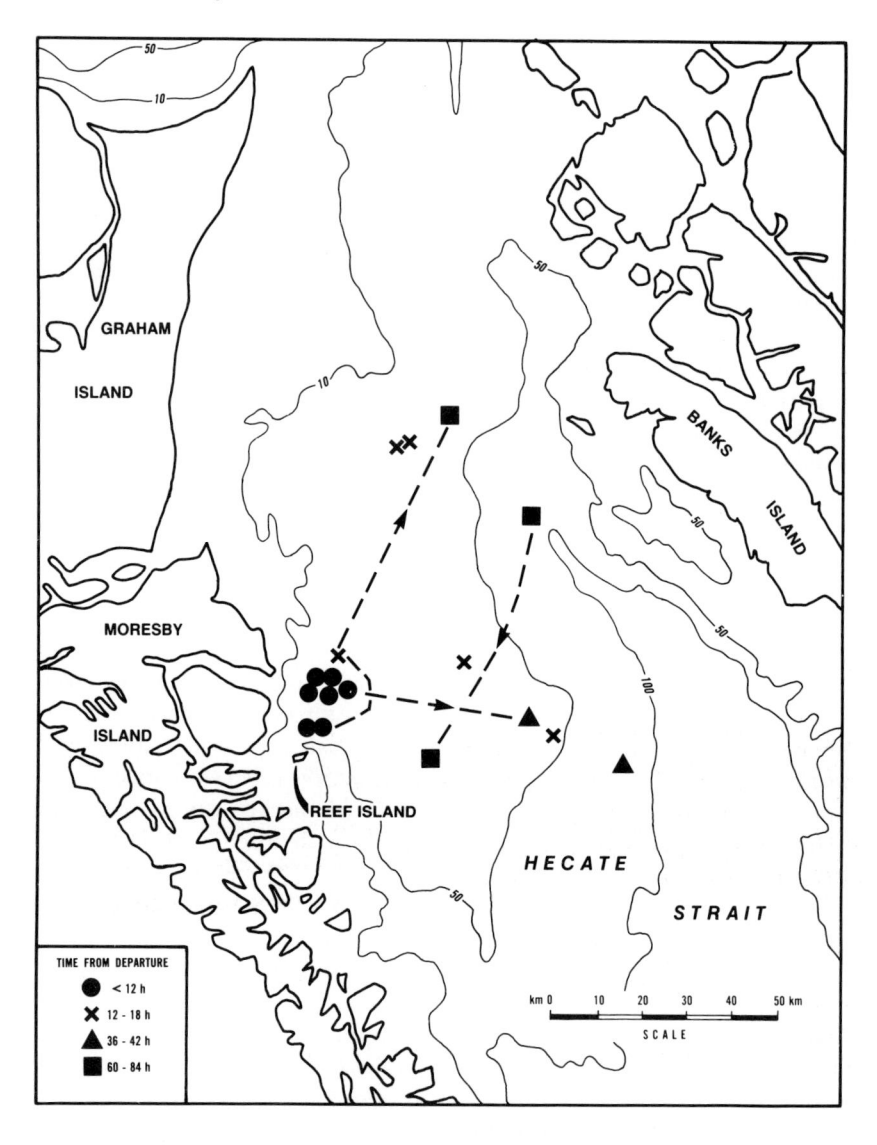

Fig. 8.15 *Positions of Ancient Murrelet broods located at sea after their departure from Reef Island in 1987.*

early July, and the timing of breeding at that colony is later than at Reef Island (Chapter 5). However, it seems probable that most successful breeders forsake the colony completely once the young leave and do not return until the following year.

FLEDGING AND INDEPENDENCE OF THE CHICKS

The information obtained by Litvinenko and Shibaev (1987) suggests that the young Ancient Murrelets are largely dependent on their parents for 6–7 weeks after departure from the colony, by which time they are more or less fully grown and capable of flight. Those chicks that they observed waited on the surface while their parents dived, then rushed over to them wherever they surfaced, to beg for food. Towards the end of the period of dependency, the adult murrelets would surface with food in their bills and then dive underwater again, apparently luring the chicks to follow them, and perhaps training them in underwater pursuit. However, even quite young chicks sometimes feed themselves, because in the Aleutian Islands Doug Forsell (pers. comm.) observed a chick aged less than one week diving to 30 cm or more to chase polychaete worms attracted to the lights of a ship at night.

The observations of Litvinenko and Shibaev agree with those of Sealy (1975), who found that young Ancient Murrelets, independent of their parents, began to appear around Langara Island about 10 July, six weeks after the first departures from the island.

CHAPTER 9

The behaviour of non-breeders

Timing of attendance; singing and calling; prospecting for burrows; what we think the non-breeders are doing.

Like other seabirds, Ancient Murrelets usually visit their breeding colonies as non-breeding prospectors for at least one season before they actually attempt to breed. There are a number of important decisions that the birds must make prior to commencing breeding; the colony, the area and the actual burrow have to be chosen, and a suitable partner has to be found. For the Ancient Murrelet, there is also the special problem of how to do all this in pitch darkness, and at the same time avoid getting caught by predators. Because prospecting birds are less attached to particular sites than breeders they are generally much harder to study. For us, at Reef Island, the question of how to distinguish non-breeders from breeders posed an immediate problem.

Principally, we used the presence, absence, or the degree of development of brood-patches, to distinguish breeding birds from non-breeders. By early May, in most years, the majority of the birds that we trapped had fully developed brood-patches. The brood-patches were covered in loose skin, which becomes highly enriched with blood during incubation, although this feature is a little difficult to see when they are caught just after arriving from the cold sea. In fact the brood-patches are difficult to find until you become experienced. On a typical passerine bird the brood-patch can be exposed by merely blowing gently beneath the abdominal feather tracts. However, the breast feathers of the Murrelets, like those of other auks, are very coarse and stiff and grow densely all over the belly, rather than being in well-separated tracts. Finding them requires running a finger forward from the region of the cloaca, a little to the side of the mid-line, to lift the feathers up and expose the brood-patches. To be sure that a bird had one, we generally repeated the procedure several times. When we found a patch we could be 100% sure of our observation, but incipient brood patches, where only a few feathers had been shed, were hard to locate, and we probably missed some of these.

Adult Ancient Murrelet on the forest floor (Photograph by Ian L. Jones).

TIMING OF ATTENDANCE

Singing and surface activity declined as the proportion of pairs incubating rose, so that by late April in most years there was a period of relatively little activity at the colony. This lasted until about the middle of May, when non-breeding birds, including some 2-year-olds banded on the island as chicks, began to arrive in large numbers. As birds rarely visited the colony in their first year (Chapter 12), many of the 2-year-olds arriving in May were presumably touching land for the first time since their departure soon after hatching. Their behaviour seemed to bear this out, as they were much more easily caught than the breeders, and seemed clumsier at making their way through the canopy. Notes from the 1989 log give some impression of events at this stage in the season:

> 20 May (light wind after several windy days, full moon after 0100, but cloudy). "Birds began to arrive while still very light and were quickly showering out of the

trees. They landed indiscriminately on all parts of the colony and many were found in dense regeneration. It was clear that most were unfamiliar with the area. Many were sitting on the ground, although it was still very light, even at 0100, because of the moon. After 0030 there was lots of singing, and after 0200 many birds were singing from the ground and [chatter calling] in burrows. Numbers continued high until 0400, by which time dawn was clearly visible and no flashlight was necessary for walking. We obtained few retraps from this 'shower' of birds—only 2 out of 47 captured [which included 38 without brood patches]."

22 May (light wind, some cloud, moon rose late). "A very big night for arrivals. Birds began to arrive at plot C by 2330 and hit the ground in large numbers after 2345. Many came in groups of 3 or 4, almost simultaneously. This continued until 0300, with many singing and sitting about on the ground."

31 May (Moderate wind, cloudy, no moon). "Very dark at 0100. Not many birds on the ground, but many singers and after 0200 many birds calling from burrows. Only 1 out of 22 trapped was a retrap. One bird caught while singing [chatter calling] in a burrow had a bill depth of 7.4 mm [>95% probability of being a male], weighed 197 g and had no brood patch."

We have no information on how frequently individual non-breeders visited the colony, but by June most seemed more adept at avoiding us, suggesting that they had gained some experience in how to behave on the colony. Burrow excavation was in evidence from mid-May onwards, and continued at least until mid-June (Figure 9.1), with many non-breeders exhibiting soil on their feet and bills. The bill measurements of birds which appeared to have engaged in burrow excavation included those characteristic of males and females, but there appeared to be a preponderance of males (24/35 had bill depths of 7 mm or greater). In two cases two birds, both with muddy bills and claws, were captured together in the same burrow, so it seems likely that both sexes sometimes undertake burrow excavation. However, males may be more prominent in this activity than females.

SINGING AND CALLING

Usually a great deal of singing and *chirrup* calling occurred on nights when large numbers of non-breeders were trapped. In addition to these vocalizations, delivered above ground, there was a lot of calling from burrows. As in the early part of the season, the first arrivals frequently perched in trees before landing on the ground. Early calling came mainly from the canopy. Later in the night, some birds gave *chirrup* calls from logs or stumps, while others gave *chatter* calls from the entrance to or from within burrows. Among birds captured on the surface at night, the proportion of non-breeders rose from 20% before midnight to more than 80% by 03.00 h (Figure 9.2). Presumably most breeders arrived early and quickly disappeared underground.

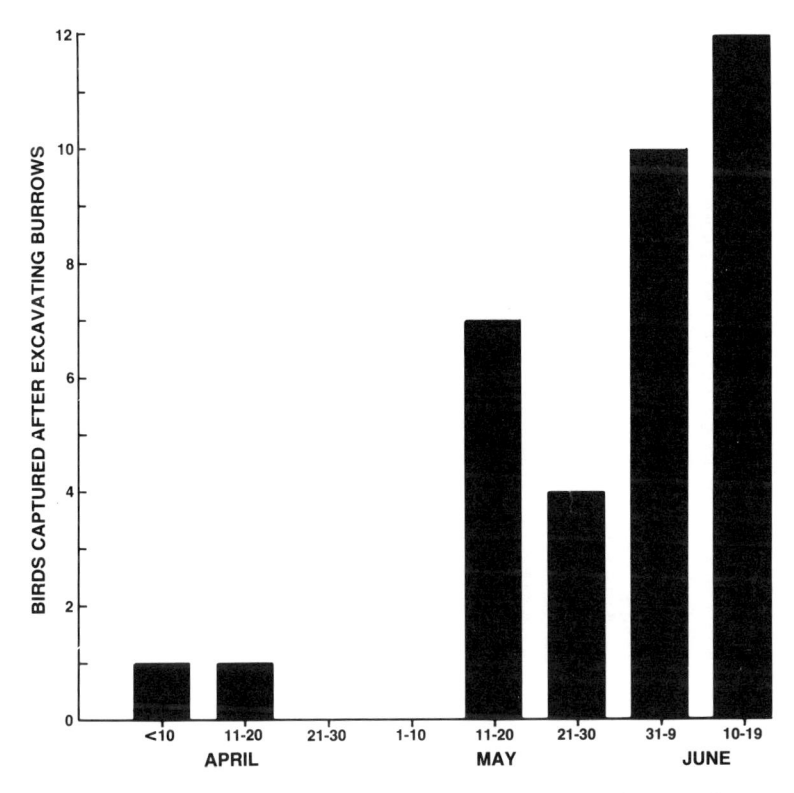

Fig. 9.1 *Numbers of birds captured at different times of the season that had been actively excavating burrows.*

If we approached a bird calling from the ground we frequently found, on switching on a flashlight, that there were several other birds crouched nearby. Likewise we often found up to four birds sitting near, or just inside, the entrance to burrows from which calling could be heard. We caught three birds in this type of situation, and their bill measurements suggested that they were females. The playback of songs and *chirrup* calls during the latter part of the season attracted Ancient Murrelets, which landed nearby, suggesting that these vocalizations serve as an advertisement by males, presumably to attract females.

Vocalizations in burrows where no breeding attempt had been made that year usually occurred two or more hours after the start of arrivals, and was most common near to the end of the night, continuing until the last birds were departing in the early morning. In contrast, vocalizations in breeding burrows occurred invariably when the off-duty member of a breeding pair returned. This could occur at any time during the night, but usually happened within the first 2 h after the start of arrivals. In mated pairs the greeting ceremony involved simultaneous calling by both birds. *Chatter* calling from non-breeding

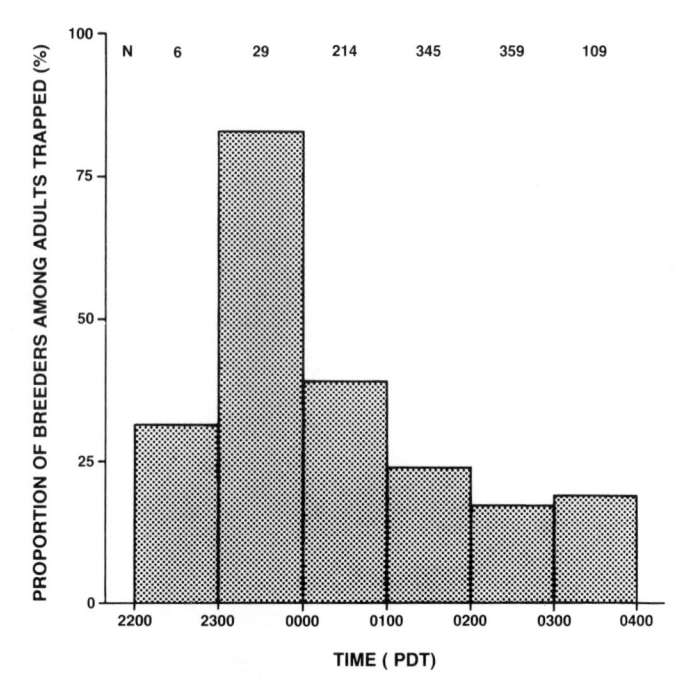

Fig. 9.2 *The proportion of breeders among birds captured at Reef Island during 21 May to 10 June, in relation to time after the start of arrivals (PDT = Pacific Daylight Time).*

burrows was generally by only one bird at a time, but sometimes, and especially towards the end of the night, two birds could be heard calling together. This may mean that they were close together, as is the case when pairs call together in the nest-cup, but we do not know that for sure.

In some cases *chatter* calling was heard from burrows which had been used that season, but from which the brood had departed. We were able to trap the two birds involved on seven occasions; none had fully developed brood patches. In five cases the bird with the deeper bill had an incipient brood patch, from 10 to 17 mm across, while the other bird was without one. One of the burrows was occupied the next year by the same breeding pair as in the previous year, while the other was unoccupied. Perhaps it is not surprising that the non-breeders did not return after I had interrupted their courtship so rudely. Three other birds caught in breeding burrows, after the breeding pair had left, were without brood patches. A fourth, a banded 4-year-old, had an incomplete brood patch.

PROSPECTING OF BURROWS

At the end of the season we found that many burrows from which families had departed were being entered regularly. In fact those that had been used were entered more often than those which had not (Figure 9.3). At first, we assumed that the visiting birds were members of the original breeding pair. However, as pointed out in Chapter 8, we obtained fairly good evidence from the seven breeders equipped with radio transmitters, and from the paucity of retraps of birds banded while departing with chicks, that few successful breeders returned to the colony, at least for several weeks after accompanying their chicks to sea. These observations, combined with the capture of non-breeders in several burrows, suggested that most birds that entered burrows after families had departed were prospectors. There was a striking difference between areas D and F in the proportion of burrows entered during the post-breeding period (Figure 9.3). At area F in 1989, an average of 50% of burrows used that year were entered each night, compared with only 13% at area D. This appeared to be the result of much larger numbers of prospectors visiting area F, a phenomenon for which we found additional evidence.

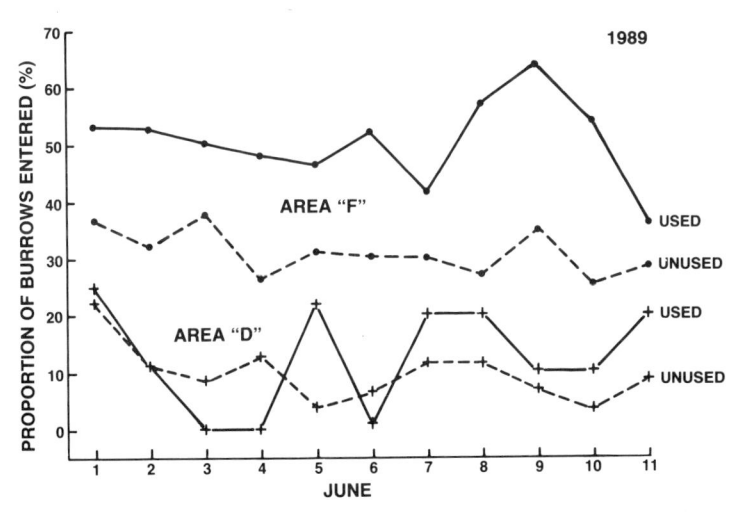

Fig. 9.3 *The proportion of burrows entered per night in relation to whether or not they had been used that year (1989).*

Birds interacting in burrows became intensely involved in their activities and frequently continued when a light was shone at them, even, in some cases, when an arm was extended into the burrow to remove them. The chattering bird sat on the ground with its bill tilted upwards, and raised the feathers of its forehead, a display also seen frequently among birds interacting on the gathering ground. The interactions were eventually terminated at dawn. We never found birds displaying in this way on the surface.

I have stood at the mouth of a burrow in the early morning as chattering continued, while the outlines of nearby trees slowly asserted themselves. Gradually, the *chatter* calls became intermittent and finally ceased. Within five minutes the birds, invariably two, but separated in their departure by one or two minutes, would shuffle to the burrow entrance, look about for a few seconds, and then launch themselves towards the sea. I caught eight birds as they emerged and all were non-breeders, although four had incipient brood patches, less than 15 mm across. The possibility that some of these birds had been members of pairs which laid, but never initiated incubation, cannot be discounted.

Departure from the colony in the early morning generally occurred closer to dawn during late May and June than earlier in the season, perhaps because the nights then are very short. In early April the timing of first arrivals and last departures were more or less symmetrical about the darkest period of the night, with birds arriving about 2 h after sunset and leaving about 2 h before dawn (Jones, 1985). By early June, when the nights were very much shorter, the last birds left as late as 1 h before sunrise, by which time we could see them easily without the aid of a flashlight, something that was never possible when they arrived.

The proportion of birds trapped as non-breeders and retrapped subsequently at Reef Island was much lower than the corresponding proportion of breeders, suggesting that non-breeders may visit several different parts of the colony in the course of a season. However, those that were retrapped were generally caught close to where they had originally been banded (Chapter 12). Probably non-breeders begin by prospecting various parts of the colony, and if caught at this stage are unlikely to be retrapped. Later they become attached to a particular area, and those that became attached to the area in which we carried out most of our trapping were the ones that we retrapped.

During May and June of 1984 we collected six birds that were singing in trees. All proved to be males without brood patches, and hence were presumably non-breeders (Jones *et al.*, 1989). Over the years, we also managed to trap 12 other birds while singing, or *chirrup* calling, on or near the ground. All except one were without brood patches, and hence were presumably non-breeders. The one bird with a full brood-patch was trapped on 11 May, the rest on various dates between 10 May and 14 June. In no case were the measurements of birds trapped while singing outside the range of those observed for males. From this evidence we deduced that most of the singing is done by males, and that, once incubation has begun, the considerable volume of song that occurs is produced mainly by non-breeders.

The influx of non-breeders in mid-May, which was apparent from numbers of birds trapped on the surface and the amount of vocalization going on in the colony, showed up only faintly in the frequency of knock-downs at burrows. Burrows in which no breeding had occurred were not entered any more often towards the end of the season than they had been earlier on (Figure 9.4). Instead, the proportion of burrows entered each night tended to decrease in areas C and D, and remained fairly stable in area F. These observations, combined with the trapping of many birds that had obviously been involved in

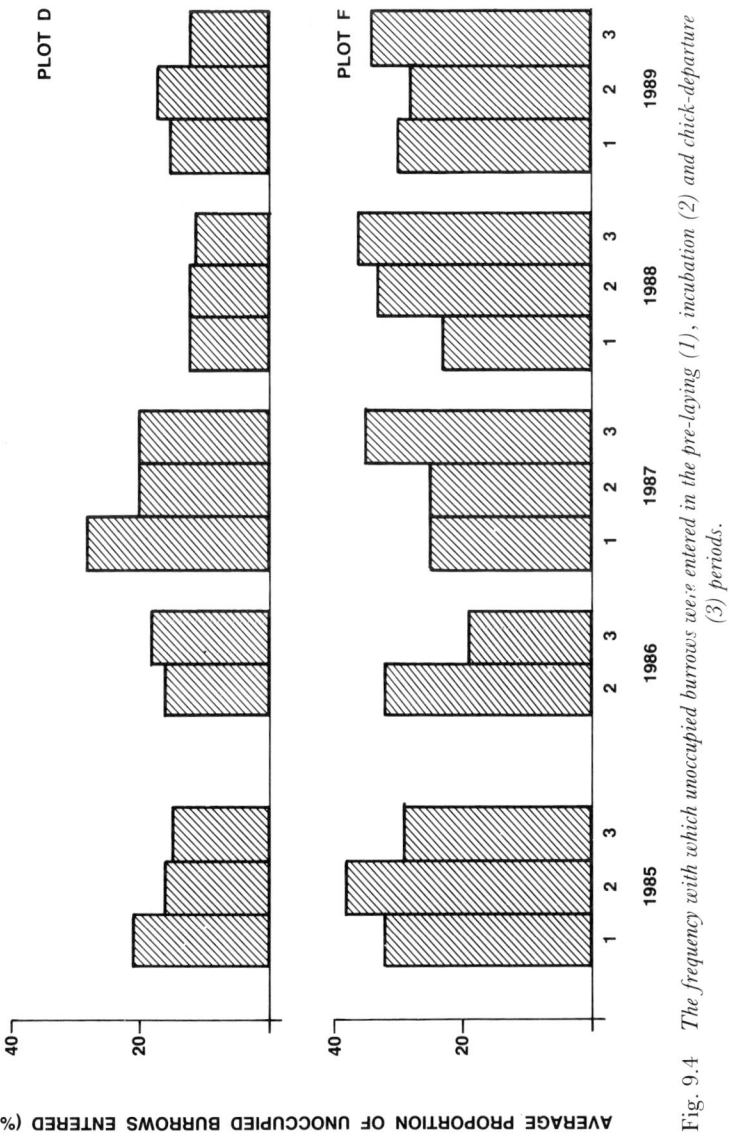

Fig. 9.4 *The frequency with which unoccupied burrows were entered in the pre-laying (1), incubation (2) and chick-departure (3) periods.*

excavation, suggest that a substantial (but unknown) proportion of the non-breeders arriving at the colony after the middle of May were engaged on digging new burrows, rather than looking to take over existing ones. Nevertheless, some burrows that had been unoccupied for one or more years were reoccupied during our study, and some others were occupied by different pairs in successive years (Chapter 12). Consequently we must assume that some prospecting birds will take over existing burrows, rather than digging new ones.

In some years, patterns of attendance during the latter part of the breeding season were similar to those seen during pre-laying, with nights of great activity interspersed with nights of very low attendance. However, in years when feeding conditions appeared to be good, particularly 1987 and 1989, large numbers of non-breeders visited the colony nightly, except when adverse weather conditions occurred.

WHAT WE THINK THE NON-BREEDERS ARE DOING

Piecing together the tantalizing clues provided by our very incomplete observations on the behaviour of non-breeders leads to the following deductions, some of which remain very speculative. During the latter half of the breeding season, male non-breeders land on the colony from about 2 h after sunset. These birds begin by calling or singing in the trees, and we have no idea why they do one or the other. However, a particular bird usually will deliver either songs or *chirrups*. Females may land in the trees at the same time, but as we have not identified any calls as unique to females, we have no way of knowing. After a time, some males begin to fly down to the ground. There, they may continue *chirrup* calling (sometimes singing), or give *bubble* calls, or they may enter burrows and begin *chatter* calling. Possibly they *chirrup* outside the burrow first, and then enter, but we have no evidence on that except that calling in burrows is most frequent at the end of the night. Both activities attract other birds, presumably females, which land nearby. Eventually, one or more females may enter the burrow and one may join the male and the pair will begin simultaneously *chatter* and *bubble* calling. This is apparently part of the process of pair formation. Activity on the surface is greatest on dark nights, and least when there is bright moonlight.

There are some puzzling features in the above scenario. The most obvious question is, if pairing always takes place in a burrow, does this mean that birds excavate burrows before pairing? The simplest explanation would be that the male digs the burrow, then entices a female to accept it, as occurs in the case of some passerine nests (e.g. Winter Wrens). However, the measurements of birds trapped after digging suggested that both sexes dug burrows. In fact, *chatter* calling was often heard coming from very large, wide-mouthed burrows, which were rarely used for breeding. Consequently, I believe that after pairing takes place, typically in the large display burrows, many pairs shift elsewhere to actually breed, either digging a new burrow themselves or finding a smaller and more attractive pre-existing burrow. If this is the case, then initially it must be some quality of the male, rather than the burrow that he occupies, that determines the choice of the female. This perhaps explains why we saw very little sign of fighting over the possession of burrows, although couples pecking at one another, or chasing about on the forest floor, were sometimes seen.

I came to think of certain large cavities, where courtship calling was audible very frequently, sometimes coming from more than one hole, as basically being display grounds (some of my colleagues referred to them as dance halls, although we never saw the murrelets dancing) where non-breeders had gathered for pairing. It is tempting to equate this idea with a lek, an area where male birds (occasionally females) gather to display, and members of the other sex congregate to compare potential mates. However, it is equally possible that these areas of intensive courtship activity simply indicated promising breeding habitat which many non-breeders were in the process of colonizing (see Chapter 12). We got to know such areas rather well, because they were good places to trap incoming birds. If the final stages of courtship display take place underground, but not in the eventual breeding burrow, then the function of the display burrow must be basically one of concealment. As I mentioned, birds involved in *chatter* calling were very unwary. On the surface they would have made easy targets for predators. By carrying out this part of the courtship in a burrow, Ancient Murrelets provide themselves with protection against predators hunting by sight. Moreover, I have found the *chatter* and *bubble* calls very difficult to locate. To the human ear, at least, the *chirrup* call provides a much better directional clue than the *chatter*. This makes sense if we consider that the *chirrup* is being used as a wide-ranging advertisement, whereas the *chatter* and *bubble* calls are designed for private communication between individuals very close to one another.

CHAPTER 10

Behaviour on the gathering ground

The position of gathering grounds relative to the colony; behaviour on the gathering ground; what birds attend the gathering grounds and what are the gathering aggregations for?

It is not uncommon to find large flocks of seabirds on the water close to their colonies, bathing, feeding, or socializing. Such aggregations tend to be ephemeral and may occur at any time of day. However, some seabirds that visit their breeding sites at night make a regular habit of gathering offshore in the late afternoon during the breeding season, before flying in to the colony. This behaviour, is very marked in certain shearwaters (Lockley, 1942; Warham 1990). Among auks, it is most obvious in the Ancient Murrelet.

THE LOCATION OF THE GATHERING GROUNDS

In the Ancient Murrelet the daily gathering congregation occurs on a well-defined area, usually within sight of and 1–3 km from the breeding area (Sealy, 1976; Vermeer *et al.*, 1985). Birds also assemble again in the early morning, before departing for the feeding area. Several of the more prominent gathering grounds in the Queen Charlotte Islands have been mapped by Rodway *et al.* (1988), from whom Figure 10.1 is derived.

At Reef Island we mapped the position of the gathering ground by conducting boat transects back and forth across the area 1–2 h before sunset, running at a constant speed on a fixed heading, and plotting the position of each group

Groups of Ancient Murrelets on the Reef Island gathering ground. Low Island is behind and to the right.

of birds estimated to be within 200 m of our course by dead reckoning. Examples of surveys made on 29 May and 1 June 1989 are shown in Figure 10.2.

We found that the gathering ground used in the evening was situated roughly in the centre of the triangle formed by Low Island, South Low Island, and the centre of the north coast of Reef Island. There was another regular gathering ground just east of East Limestone Island, and a smaller one, perhaps not always used, between South Low and East Limestone islands (Figure 10.3). In the early morning the gathering ground appeared to be more spread out, with many birds found well to the east of the main evening concentration.

WHAT BIRDS ATTEND THE GATHERING GROUND?

Although the breeding population of Reef Island is three to five times that of Limestone Island, we found higher densities on the gathering ground adjacent to Limestone Island (average of five surveys, 139 birds/km^2) than on the one off Reef Island (average of seven surveys 73 birds/km^2) in 1989 (Table 10.1). It may be that birds bound for either of the two colonies use these two areas indiscriminantly. In the area off Limestone Island (21.1 km^2) we estimated 2928 (range 2252–4109) birds on the water, assuming that we counted all birds within 150 m of our course (although we counted birds up to 200 m from our course, we probably did not see all the birds at the edge of the transect strip).

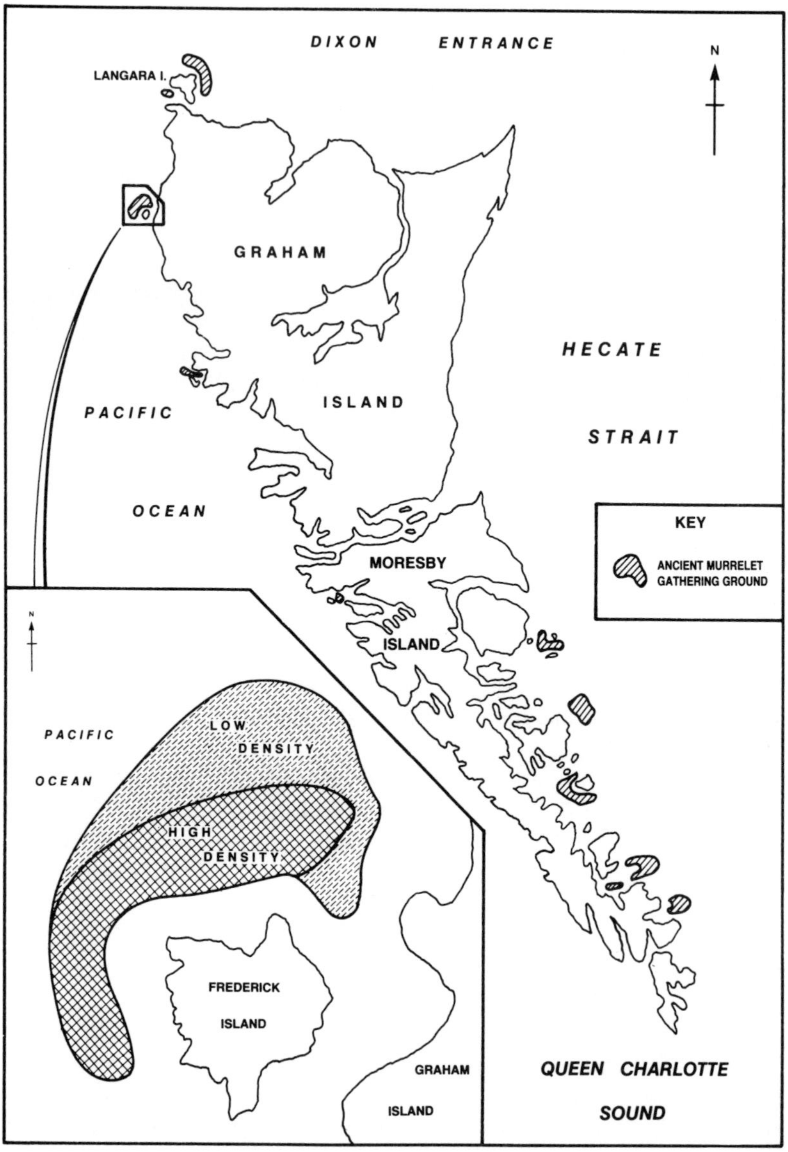

Fig. 10.1 *Position of Ancient Murrelet gathering grounds in the Queen Charlotte Islands (after Rodway et al., 1988).*

The corresponding estimate for the Reef Island area (20.4 km^2) suggested about 1493 birds (range 444–2946).

Surveys carried out during the period when most breeding pairs were still incubating suggested an average attendance at the two gathering grounds

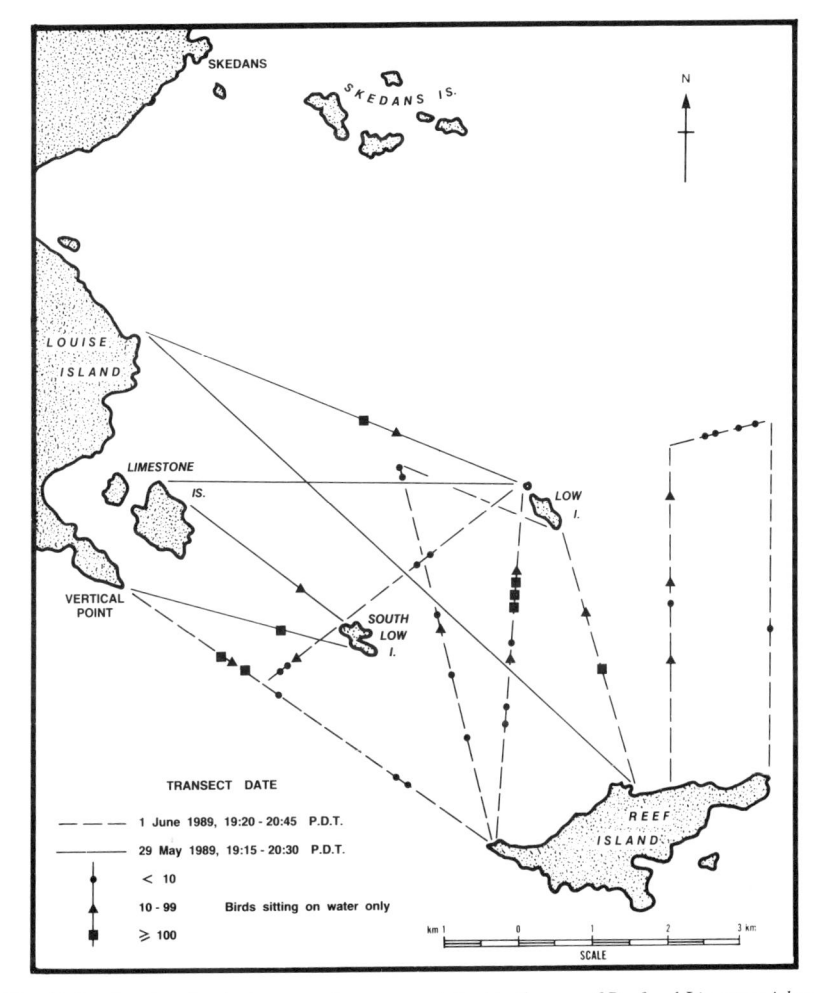

Fig. 10.2 *Results of gathering ground surveys carried out in the area of Reef and Limestone islands on 29 May and 1 June 1989.*

combined of about 4000 birds (Table 10.1). Reef and the Limestone islands between them support approximately 6000 breeding pairs, and between a half and one third of these should have been visiting the colony on any one night (Chapter 8), so there could have been 2000–3000 returning breeders present on the gathering ground. The balance could have to be made up of non-breeders. However, it is clear from our counts that, on some nights, the majority of the gathering ground flock must be made up of non-breeders.

On 29 May 1989 we estimated more than 3000 birds on the Limestone Island gathering ground, and on 1 June, about 3000 on the Reef Island area. By that stage of the season more than three quarters of family parties had left

146 *Studies at Reef Island*

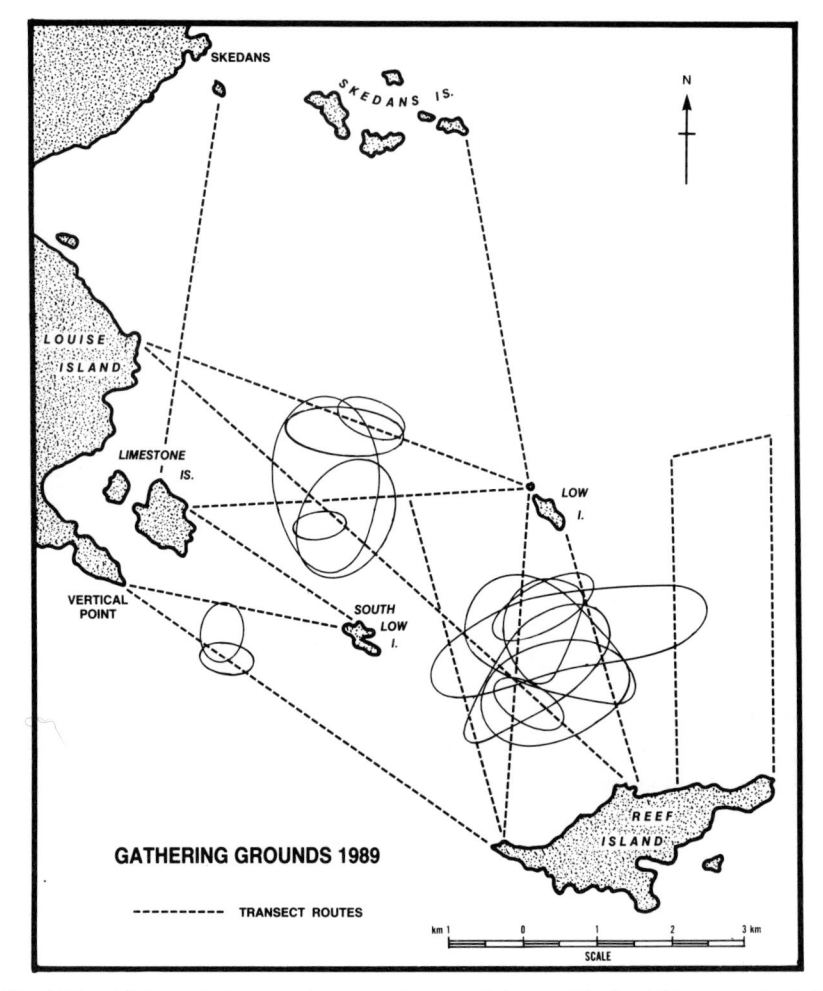

Fig. 10.3 *Major gathering ground concentrations recorded around Reef and Limestone islands in 1989.*

the vicinity of the colony, and consequently less than 1000 breeders should have been attending the Reef and Limestone island colonies nightly. Hence, we must presume that the majority of birds on the gathering grounds were nonbreeders. The counts indicate that sufficient non-breeders were present in the area to account for nearly all the birds seen on the gathering ground during the incubation period. We do not know whether the majority of birds attending the gathering ground in late April were breeders or non-breeders. However, I think that they were mainly breeders, because few non-breeders came ashore at that date.

Throughout the season there was a strong correlation between the number

Table 10.1 *Counts of Ancient Murrelets on boat transects of the gathering grounds adjacent to Reef and Limestone islands in 1989.*

Date	Distance covered (km)	Murrelets seen	Density (Birds/km^2)	Estimated No. present
REEF ISLAND AREA				
24 March	22.7	483	70.8	1444
7 April	29.9	921	102.7	2096
17 April	32.6	213	21.8	444
4 May	32.6	295	30.1	615
20 May	29.9	924	103.0	2103
22 May	32.6	385	39.3	802
1 June	33.1	1434	144.3	2946
LIMESTONE ISLAND AREA				
4 April	17.5	563	107.5	2265
11 April	22.0	839	127.1	2678
28 April	23.3	1365	195.0	4109
14 May	23.3	748	106.9	2252
29 May	23.2	1102	158.4	3337

of birds visiting the gathering ground and the numbers arriving on the slopes that night (Figure 10.4). This also suggests that the number of birds on the gathering ground was determined mainly by the presence of non-breeders, because we found that numbers of birds visiting the colony correlated much better with entries into unoccupied burrows than with incubation changeovers at occupied ones. Both counts were affected by weather conditions

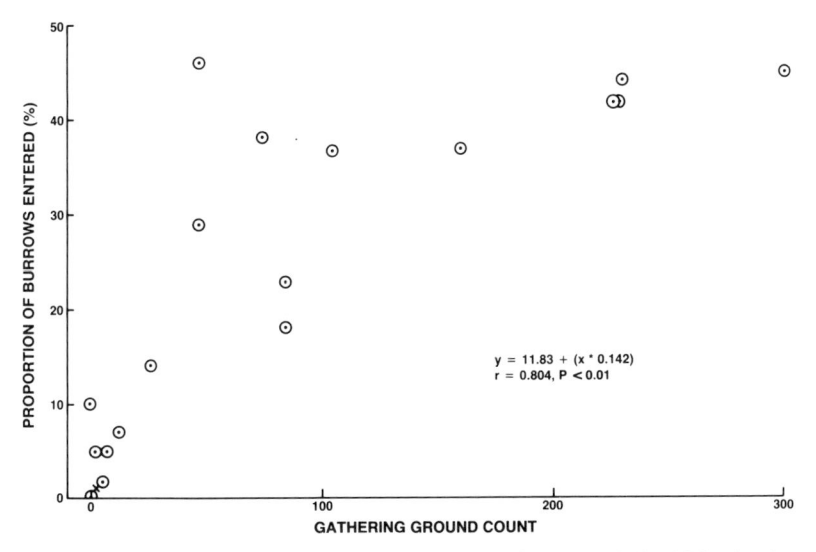

Fig. 10.4 *The relationship between evening counts of birds flying over the Reef Island gathering ground, and the proportion of burrows entered that night (1989).*

$y = 11.83 + (x * 0.142)$
$r = 0.804, P < 0.01$

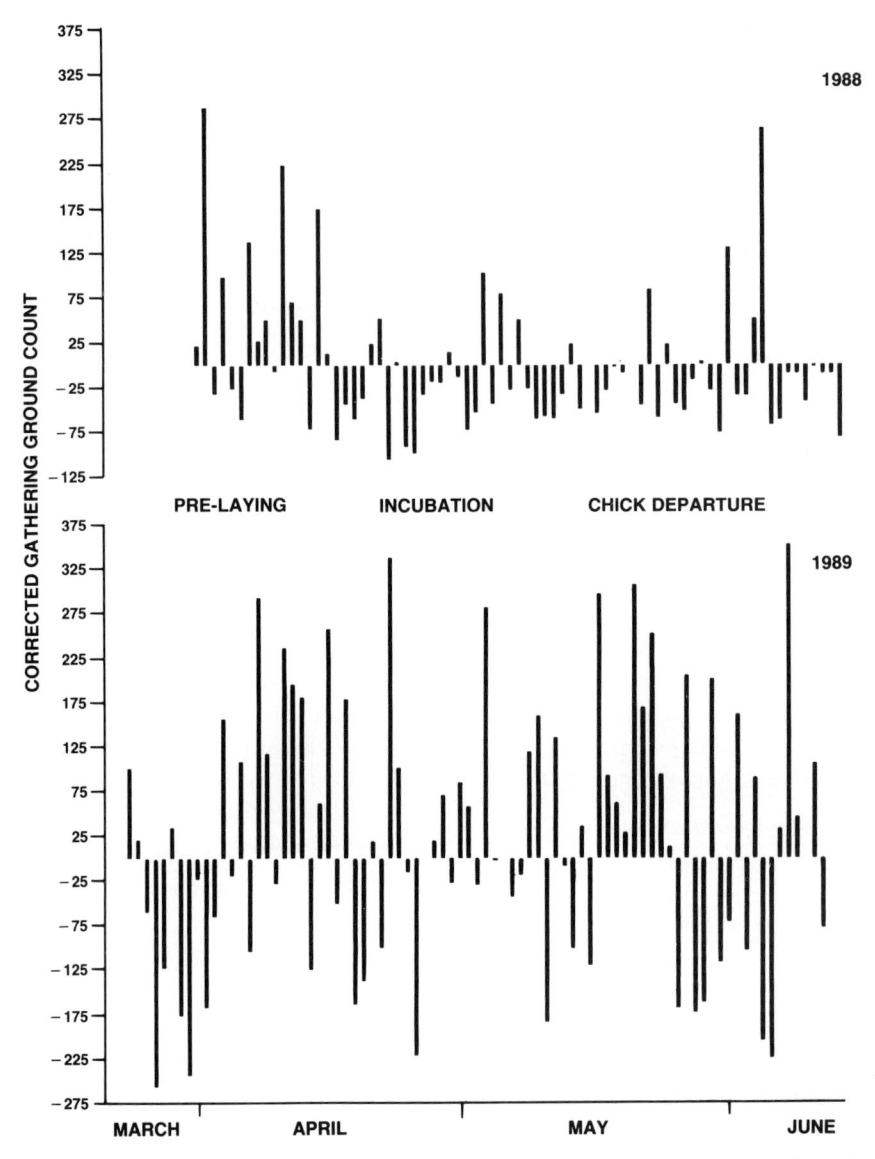

Fig. 10.5 *Corrected gathering ground counts (deviation of count from regression on wind speed) made from Reef Island in 1988 and 1989.*

(Jones *et al.*, 1990), especially wind speed and sea conditions, which were highly correlated. We eliminated the effects of wind on the number of birds on the gathering ground by calculating the difference between the actual number seen and the number predicted by the regression of the count on wind speed. When we made this correction we found that in 1988 counts tended to be

Behaviour on the gathering ground 149

highest during the first half of April and lowest in late April and the first half of May. The period of low counts corresponds with the period of low activity prior to the arrival of the majority of prospecting nonbreeders. In 1989 there was no clear pattern, although counts were again high in the first half of April (Figure 10.5). It is possible that in 1989, an earlier year than 1988 (Chapter 13), non-breeders began to visit the gathering ground in large numbers earlier in the season.

In addition to our daily inspection of the gathering ground, 2 h before sunset (see above), we also periodically performed the same 10 minute count every hour from mid-day onwards throughout the season, on days when weather conditions appeared suitable. This allowed us to observe the build-up of birds on the gathering ground, at least to the point about half an hour before sunset when light conditions made counting impossible. Birds generally began arriving on the gathering ground between 16.00 and 18.00 h throughout the season, so the length of time spent between arrival on the gathering ground and landing on the colony increased during the season (Figure 10.6). Peak counts on several nights in the second half of the season occurred 1 or 2 h before the final count, suggesting that most birds arrived on the gathering ground well before dark.

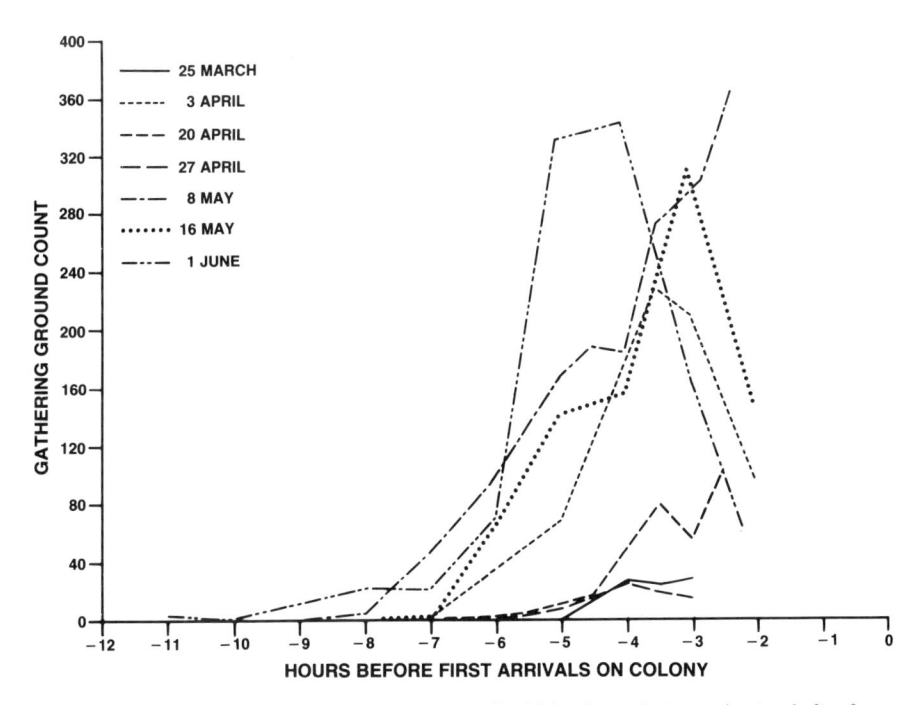

Fig. 10.6 *Build up of gathering ground counts at Reef Island, in relation to the time before first arrivals at the colony.*

Ancient Murrelets displaying on the gathering ground off Limestone Island (Photograph by Colin French).

BEHAVIOUR ON THE GATHERING GROUND

On calm evenings we watched the behaviour of birds on the gathering ground opposite our camp through a telescope. We also visited the gathering ground periodically by boat to watch the birds and listen to them calling. The most striking feature of the gathering ground was the amount of activity going on; the birds were not simply waiting for darkness, they were very actively engaged in socializing. The most common vocalization given on the water was the *chirrup* call, and at a distance the sound coming from the flock resembled that from a roost of sparrows, making their Japanese name, "sea sparrow", particularly appropriate. Some birds seemed to be in pairs, circling around one another and calling, with the feathers of the forehead raised, and trios were also common. The majority of birds formed loose flocks, often spread out in long lines. These flocks constantly formed and reformed, as birds shifted from one part of the gathering ground to another, and individuals were constantly taking off to fly short distances before pitching again on the water. These brief flights were often characterized by an unusual amount of swerving and banking from side to side, and some clearly functioned as displays.

A very characteristic display, seen often on the gathering ground, we termed the "flop display". A bird would jump clear of the water, as though taking off, and then plunge back from about 30 cm above the surface, usually turning as it did so, so that it dived in sideways with wings spread, almost as though cartwheeling. The flop display was clearly described by Shibaev in Flint and Golovkin (1990), although he considered it to represent an aborted take-off. This display was contagious. After one member of a group had performed it there were usually a rash of other flop displays within a few seconds. It was often performed by one, or occasionally both, members of an apparent pair, and among trios two birds often performed it one after another. The pattern of one bird in a pair, or of two birds in a trio, performing the flop display suggests that it may be a male display.

Not much diving occurred on the gathering ground, although sometimes one bird, a pair or a group would submerge briefly. In comparison with murres, which do a lot of diving while socializing on the water (pers. obs.), Ancient Murrelets seem to do relatively little.

Behaviour on the gathering ground in the early morning is similar to that in the evening, but we made fewer observations at that time of day. Many birds were seen flying eastwards into Hecate Strait from sunrise onwards. On 2 June 1989 we estimated over 1000 within 3 km NE of the island at 06.00 h (2 h after sunrise), and on some days several hundred remained as late as 4 h after sunrise, mainly in an area about 3 km northeast of Reef Island. At Hippa Island, on 28 May 1983, Mike Rodway, Moira Lemon and I estimated 53 000 murrelets on the gathering ground between Hippa Island and the main Graham Island at 07.00–09.00 h, about 4 h after dawn. This was at the peak period of chick departures at that colony, which supports about 40 000 breeding pairs. Such a large concentration on the gathering ground, so late in the day, was unusual at Reef Island.

WHAT ARE THE GATHERING AGGREGATIONS FOR?

The function of the gatherings that take place at the gathering ground is not very obvious. Previous writers have assumed that most birds seen on the gathering ground were breeders waiting to relieve their mates that night (Sealy, 1976; Vermeer et al., 1985). Breeders made up 75% ($n = 75$) of birds collected by Sealy on the gathering ground, but he gave no indication of the time of year, and presumably some of his collections were made before most pairs were incubating. Sealy described the behaviour that he saw on the gathering ground as "play", while Vermeer et al. noted that, off Langara Island, numbers were highest during the period of chick departures. They assumed, from this, that most of the birds were gathered "in anticipation of the hatching of their chicks". The assumption that most of the birds on the gathering ground were breeders waiting to visit the colony to exchange incubation duty was used by Nelson and Myres (1976) to obtain a rough estimate of the number of breeders on Langara Island, based on the idea that a third of off-duty birds would be gathered off-shore each night.

Any interpretation of the function of the gathering aggregations is much handicapped if we do not know for certain whether the birds that occur on the gathering ground are breeders or non-breeders. We did not collect any birds on the gathering grounds, so we have to rely on inference. I have pointed to several pieces of evidence which suggest that, once most breeding pairs are incubating, the bulk of the birds on the gathering ground are non-breeders, or failed breeders. Mike Rodway (pers. comm.) had a similar impression of Rhinoceros Auklets gathering around Triangle Island about the date of hatching.

The large amount of display behaviour exhibited by Ancient Murrelets in the gathering ground suggests that both sexes are represented, and the many birds in twos suggest that pairs were either being formed or reformed. As behaviour on the gathering ground remained essentially similar throughout the season, it seems reasonable to suppose that it functions partly as a display area. During the pre-laying and laying period, when few non-breeding birds visit the colony, most of those on the gathering ground are presumably going to breed that year. Later in the season, breeders waiting to exchange incubation duty that night may visit the gathering ground, but it is less obvious why they should do so, especially if they run the risk of being caught by peregrines in the process. The two radio-tracked breeders that were waiting for their chicks to depart did not pause on the gathering ground, but arrived direct from an area beyond the range of the transmitters. It is quite possible that, from the time that most pairs begin incubation, most of the birds on the gathering ground are non-breeders. If so, the lack of non-breeders visiting the colony before the middle of May suggests that they attend the gathering ground for some time before they begin to land on the colony.

The fact that males seem to give most of the vocalizations on the colony suggests that, as in most birds, they are the more active sex in courtship. Hence, they may be responsible for most of the flop displays. Flop displays one after another by two members of a trio may indicate two males displaying to the same female. Two birds flopping together, as a pair, suggests that both sexes may sometimes perform the display, but it is also possible that these, rather rare instances, involved two males.

One thing seems certain, the considerable amount of time and energy devoted to social behaviour on the gathering ground must have some function. The majority of auks visit their colonies in daylight, and all spend substantial amounts of time there displaying and otherwise interacting. Circumstances are much more constraining for Ancient Murrelets, which are liable to have their above-ground behaviour on the colony cut short by predators. This may provide a strong incentive for them to perform much of their courtship at sea. It seems almost certain that the importance of the gathering ground, like the strictly nocturnal timing of colony visits, derives from the danger of predation on land.

If pair bonds are being initiated or strengthened through contact on the gathering ground, then it must be important for pairs to be able to relocate one another after they land on the colony. Indeed, some arrivals at the colony involved two birds landing simultaneously very close together, and these may have been pairs which had maintained contact while flying in from the sea. Sometimes such pairs could be heard calling (a low "*chip*") as they flew in, possibly to maintain contact with one another. However, the majority of arrivals were solitary, and this suggests that it may be hard for the murrelets to keep in touch while flying in the dark. If this is so, it may provide an explanation for the loud, and persistent vocalizations that are such an outstanding feature of the colony at night. Females that have been attracted by a male's display on the gathering ground may locate him again on the colony through recognition of his call, or song. It seems a rather cumbersome system, but any other explanation for events appears even less likely.

The flop display

CHAPTER 11

Breeding habitat and burrows

The nest site; nest chamber and cup; measures taken to conceal the burrow entrance; the population of Ancient Murrelets on Reef Island; variation in burrow density and burrow occupancy.

THE NEST-SITE

The majority of Ancient Murrelet burrows on Reef Island are situated among the roots of trees, or old stumps, or under fallen logs (Figure 11.1). Those that are not protected in this way generally lead in among crevices in rocks or cavities among boulders. Burrows made in open ground and surrounded only by soil are rare. Because of this, their shape is partly determined by the character of the associated structures. For instance, stresses caused by the wind blowing the upper part of the tree tend to loosen the soil below the large lateral roots, and many burrows are tunnelled back below major roots. Likewise, the undersurfaces of fallen logs, being in contact with the damp ground, tend to rot away, forming hollow cavities beneath. Such cavities are sometimes used for burrows. As they rot away and enlarge, the spaces underneath become suitable for display areas.

Ancient Murrelets are particular about the size of their burrow entrance. Most do not exceed 15 cm maximum diameter (Figure 11.2). The length is variable (Figure 11.3), but is not usually less than 30 cm and rarely more than

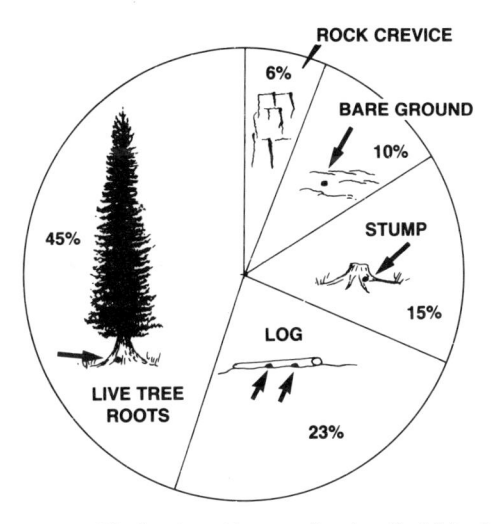

Fig. 11.1 *The location of burrows found on Reef Island.*

120 cm. Many burrows, especially the shorter ones, have a sharp bend, more often right-handed than left-handed (Figure 11.4), so that the nest chamber is completely invisible from the entrance. Occasionally birds use shallow depressions in which they are clearly visible from the outside. All such sites on Reef Island were deserted almost as soon as we found them, although we have no way of knowing whether they would have been deserted without our presence. However, birds making use of such sites would certainly have been vulnerable to day-hunting ravens, crows and eagles.

On Reef Island, and throughout most of the Queen Charlotte Islands,

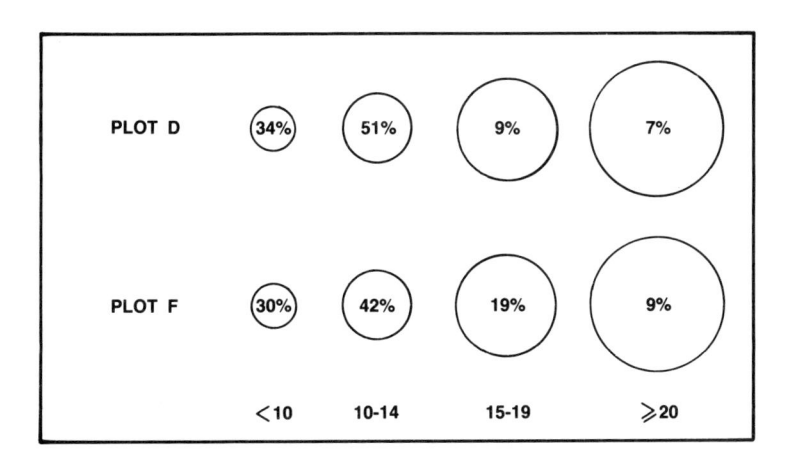

Fig. 11.2 *Entrance diameters of Ancient Murrelet burrows at Reef Island.*

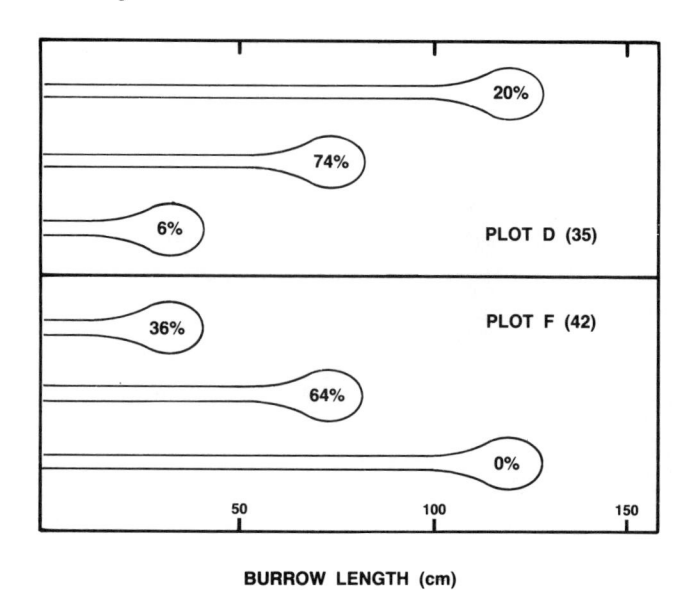

Fig. 11.3 *The length of Ancient Murrelet burrows on Reef Island.*

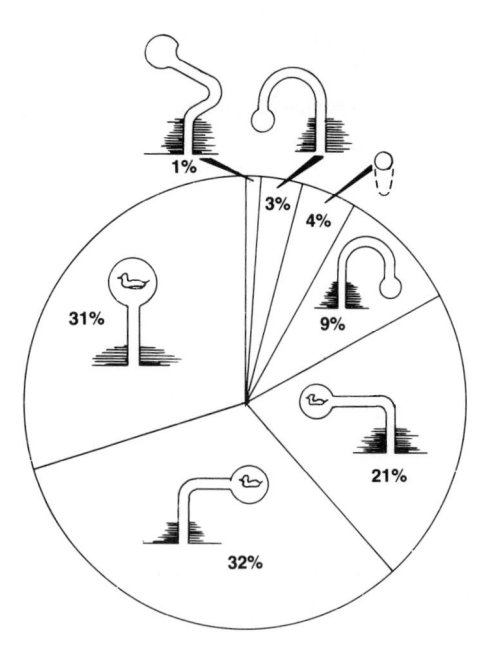

Fig. 11.4 *Shapes of Ancient Murrelet burrows at Reef Island.*

Ancient Murrelets nest very close to Cassin's Auklets. However, there is little overlap in breeding habitat. The auklets burrow almost exclusively among grassy turf within 100 m of the sea, and avoid forest, using it only where it is fairly open. In contrast, the murrelets occupy closed-canopy forest without any ground layer. On Frederick Island, Vermeer and Lemon (1986) examined the breeding habitat of the two species in detail and found that, as at Reef Island, most Ancient Murrelet burrows were under stumps, roots or logs, under forest canopy, while most Cassin's Auklets burrowed in grasslands. They concluded that competition between the two species for burrows was unlikely under most circumstances.

Yves Turcotte weighing a murrelet's egg, at the mouth of a burrow.

THE NEST-CHAMBER AND CUP

The burrow ends in an enlarged nest-chamber, varying in size from a cavity little larger than the adult bird to one as much as 50 cm across. On Reef Island the height of the ceiling varied from 7 to about 30 cm above the floor. It was not unusual for a single burrow to have two or more nest-chambers, and a single chamber might contain several nest-cups. However, it was rare to find two pairs sharing a single burrow, and we never saw more than one clutch in a nest-chamber, although some were certainly large enough to accommodate two birds. In two cases where two clutches were laid in the same burrow, but in different chambers, one of each pair of clutches was never incubated. In several other cases two burrows shared a common entrance but diverged immediately without any common tunnel. In this situation broods were reared successfully in both burrows on several occasions. Burrows with more than one entrance were rare, and where they occurred it was evident from our tags that only one was ever used in a given year.

The nest-cup usually contained fragments of grass, twigs, spruce cones and other surface litter, dragged down by the birds. I was initially in some doubt about whether this was a deliberate process, or the accidental result of the birds shuffling into the burrow. At some sites we found that the plastic tags that we placed in the entrance disappeared regularly and had to be replaced. When we examined the nest-chambers at the end of the season, we extracted as many as a dozen tags from some, neatly arranged within the cup. It seemed a rather unsuitable lining, but would have given the pairs involved a sense of being up to the minute with their interior decor. There also seemed to be more leafy twigs of hemlock in the nest than we might have expected by chance. I concluded that the birds probably made some definite attempt to select and transport nest material, although their efforts were meagre, because some nest-cups were almost bare. Frequently, there were one or two feathers in the nest chamber, but no attempt was made to incorporate these into the nest. Sealy (1976) recorded the murrelets accumulating salal and hemlock leaves in their nests at Langara Island.

MEASURES TAKEN TO CONCEAL THE BURROW ENTRANCE

Puffins and Rhinoceros Auklets kick soil out of the burrow entrance as they dig, leaving clear evidence of their activities. However, Ancient Murrelets do not do this and there is rarely any evidence on the surface that a new burrow is being dug. The ground in most places where they burrow is very soft; a person can push their hand into the soil easily. The action used by the murrelets in digging the burrow appears to involve compaction, rather than excavation. If you dig an access tunnel from the side you can usually tell that you are about to break into the burrow, because the soil becomes drier and more dense. The selection of sites where natural cracks and cavities are formed around the bases of trees or boulders further helps to disguise burrow entrances. Where the burrow leads steeply upwards, the soil at the entrance usually shows evidence of their claws. Where the ground is level or downward-sloping (67% of burrows at Reef Island), the trained eye can detect a smoothing of the hearth, but there is rarely any obvious sign of the birds' passage.

Nor do the murrelets defecate anywhere around the burrow, as Cassin's Auklets do. The burrows of the two species can usually be distinguished by the presence of white streaking at the entrance to auklet burrows, which also have a strong odour attached to them. Incubating murrelets remain without defecating throughout incubation shifts of up to five days. I have never been able to detect any smell attached to active burrows. However, murrelets removed from their burrow while incubating often release a very large volume of yellowish, liquid faeces. These have a strong, distinctive smell, reminiscent of the droppings left by a female Common Eider flushed from its nest. This is presumably a panic reaction. Very few birds trapped on the surface defecate when handled. Those few that do so defecate copiously. They are probably birds that have been relieved of incubation that night and caught while departing. Presumably incubating birds empty their gut thoroughly before arrival at the colony and the liquid faeces discharged when they are disturbed must represent mainly products of the liver and kidneys, rather than material that has only passed through the digestive tract. Denser, whiter droppings found occasionally around the bases of trees, or on logs and stumps, may be deposited by displaying birds not associated with burrows.

In a very short burrow, the birds sometimes pull twigs and other debris into the entrance behind them, so that it is partially blocked. We only stumbled on this habit by accident, having earlier ignored such evidently unoccupied burrows. Cobwebs sometimes appear across the burrow entrance during long incubation shifts, making it appear even less likely that the burrow is occupied. To be sure of finding all occupied nests every hole must be searched.

POPULATION ON REEF ISLAND

In 1985 we estimated the population of Ancient Murrelets on Reef Island, using methods similar to those of Rodway *et al.* (1988). Census plots 100 m^2 in area, were spaced at 30 m intervals along parallel transects 200 m apart over the whole of the colony. We found a mean density of 199 ± 30 burrows/ha, and a total colony area of 39.5 ha, suggesting a total of 7845 ± 1185 (S.E.) burrows on the whole island. Our sample plots covered 1.8% of the colony area. From an observed occupancy rate of 63% (see below) the breeding population was estimated to be approximately 5000 pairs.

In 1989, we censused Reef Island again, using a greater effort and a more even dispersion of census plots, but concentrating especially on the area from which chicks trapped in our catching funnels originated (Figure 11.5). We were able to define the "catchment area" of the funnels fairly accurately by observing which chicks banded on the upper slopes appeared in the funnels; it covered about 6.15 ha.

Starting from a baseline transect running through the camp, transects were placed parallel at 50 m intervals for 300 m east and 200 m west. This area included all of the catchment area of the chick funnels. The remainder of the colony area was covered by transects spaced at 100 m intervals. Census plots covering 100 m^2 were placed at 50 m intervals beginnning alternately 5 and 30 m inland from the seaward edge of the terrestrial vegetation. As in 1985, transects were continued until a thorough investigation of the area ahead revealed that there were no further signs of burrows.

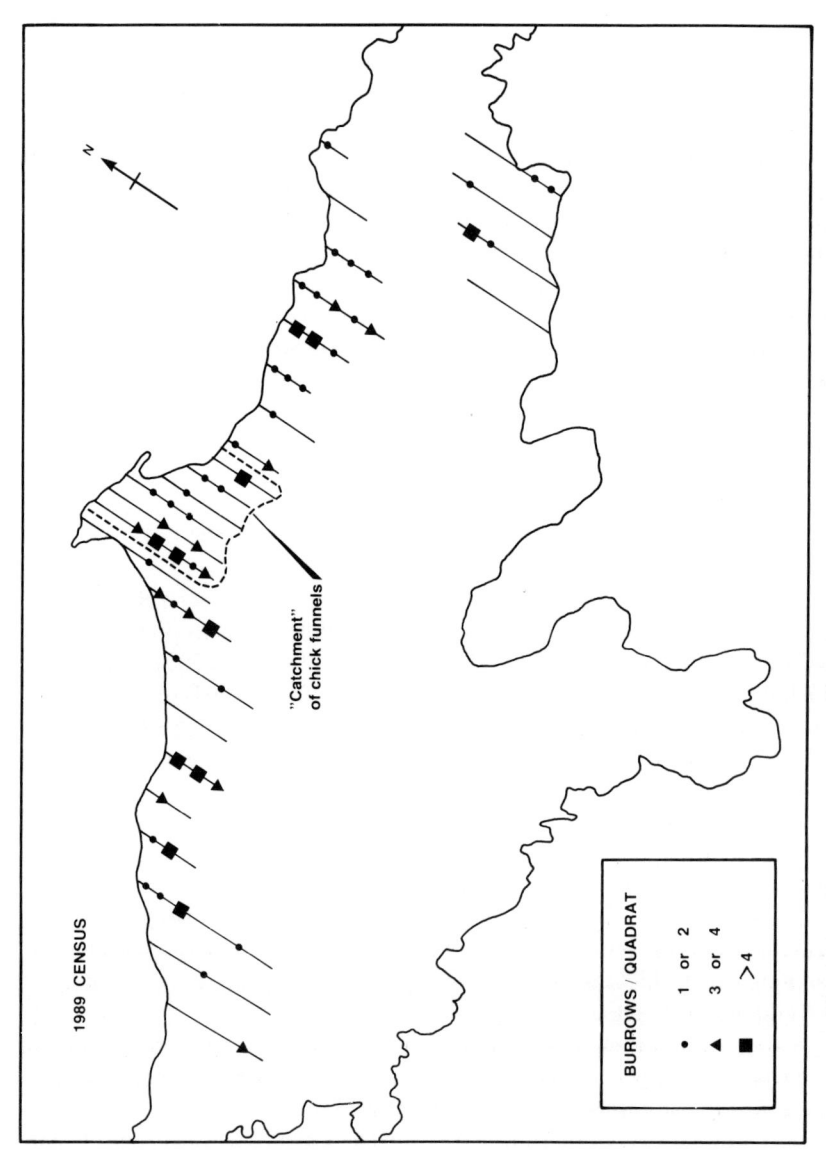

Fig. 11.5 Map showing the distribution of burrows found during the 1989 census of Reef Island.

Each census plot was searched thoroughly for burrows, which we identified on the basis of evidence such as egg-shell fragments, pieces of egg membrane, the presence of a nest-cup, or a worn appearance to the entrance. Inevitably there was some subjectivity involved when we could not reach the end of the burrow (e.g. when it went under the roots of a tree). However, by 1989 we were very familiar with the appearance of Ancient Murrelet burrows. We did not try to decide whether or not the burrow was used that year, beyond the investigation necessary to decide whether each hole constituted a burrow.

We estimated that the area for which transects were run at 50 m intervals contained approximately 1200 burrows, and the rest of the colony approximately 3900, suggesting about 5000 burrows on the colony altogether, based on examination of 2.6% of the colony area (Table 11.1). This was fewer than we estimated in 1985. However, the difference between the mean densities of burrows in the two years was not statistically significant. In addition, the number of chicks trapped in our catching funnels tended to increase during the period of the study, so it seems unlikely that the colony had declined between 1985 and 1989. Presumably the difference between the estimates arose by chance.

Table 11.1 *Estimate of numbers of Ancient Murrelet burrows on Reef Island, 1989.*

Area	Extent (ha)	Plots	Burrows	Density (burrows/ha)	Total
50 m transects	11.6	44	61	138.6 ± 28.7	1162
100 m transects	35.7	81	89	109.9 ± 20.6	3919
Totals	47.3	125	150		5081

Assuming an occupancy rate in 1989 similar to that in 1985 gives an estimate of about 3200 breeding pairs. Such estimates are subject to many possible errors, some of which (e.g. occupancy) are hard to measure. However, we could compare our estimate of burrows in the area above the funnels with the numbers of chicks trapped. Within the catchment area of the chick-trapping funnels we estimated approximately 850 burrows, 17% of the entire colony. Breeding success on the colony in 1989 was estimated to be 1.58 chicks/pair at departure from the burrow (Chapter 14). Using an occupancy rate of 63% this predicts a total production of 846 chicks from the catchment area. We actually captured 900 chicks in the funnels in 1989, which suggests that our census did not greatly underestimate the number of birds breeding within the catchment area. It seems best to regard the population of Reef Island during the period of our study as between 3000–5000 breeding pairs.

VARIATION IN BURROW DENSITY

Our census records showed that the density of burrows varied greatly within the boundaries of the colony. Out of 72 plots examined in 1985, 26 (36%) had no burrows at all, but the rest contained up to 12 burrows within the 100 m^2; a maximum density of 1200 burrows/ha. Fifty percent of burrows found were in plots containing five or more burrows (density 500/ha). Results in 1989 also revealed considerable variation among plots, with 69 (55%) of 125 census plots having no burrows, while others had up to eight burrows. Plots containing five or more burrows accounted for 44% of the burrows found. Hence about half the birds were breeding in areas where the local burrow density was 500/ha, or more.

In 1985 we recorded the extent of ground, shrub and canopy cover for each census plot, estimating them by eye. We counted the number of trees more than 30 cm in diameter at breast height and recorded their species, and also recorded the dominant shrub, and the dominant type of ground cover (moss, litter, grass, etc.). Surprisingly, we found no relationship between any of these variables and the number of burrows found in the plots. It appears that, within the colony boundary, the type and extent of vegetation has relatively little affect on the birds' choice of burrow sites. However, all of the plots had at least 30% canopy cover, most more than 50%. Forest cover is certainly an important prerequisite for Ancient Murrelet breeding in the Queen Charlotte Islands.

There was strong evidence that burrow density was affected by two physical characteristics; slope and distance from the sea. Burrow densities were much lower where the slope was less than 30° than in steeper areas (Figure 11.6).

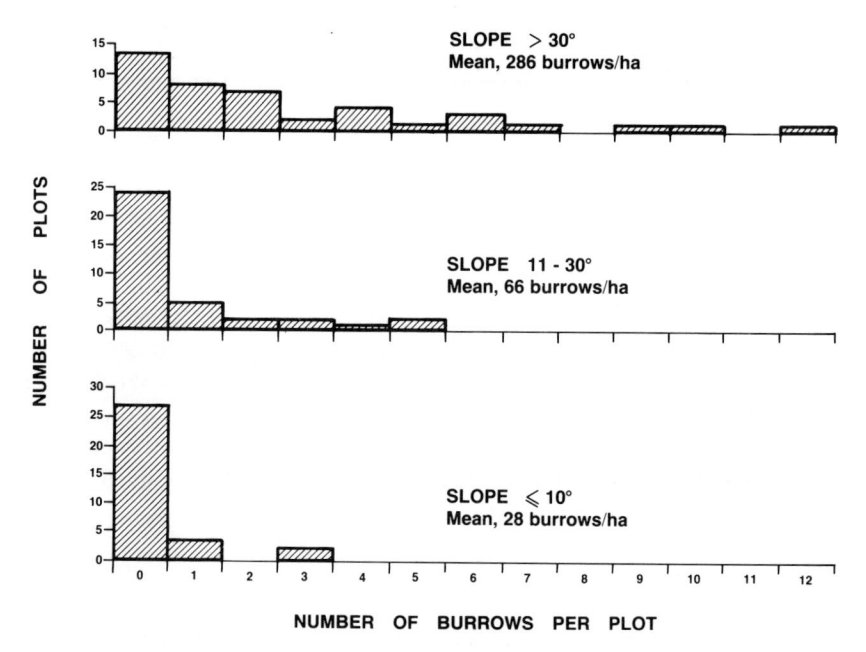

Fig. 11.6 *Burrows found per plot, in relation to slope.*

Peak burrow densities occurred about 100 m from the sea, with lower densities closer to the shore, and very low densities more than 205 m inland (Figure 11.7). There was some correlation between the slope of a plot and its distance from the shore, because the steepest areas were mainly 50–100 m behind the shore. However, an analysis only including plots between 100–200 m inland showed a similar effect of slope; plots sloping at 10°, or less ($n = 11$) had an average of 54 burrows/ha, those with slopes of 11–30° ($n = 17$) averaged 71 burrows/ha, while those on steeper slopes ($n = 17$) averaged 271 burrows/ha. Apparently the two factors both have an influence on the murrelets' choice of breeding area.

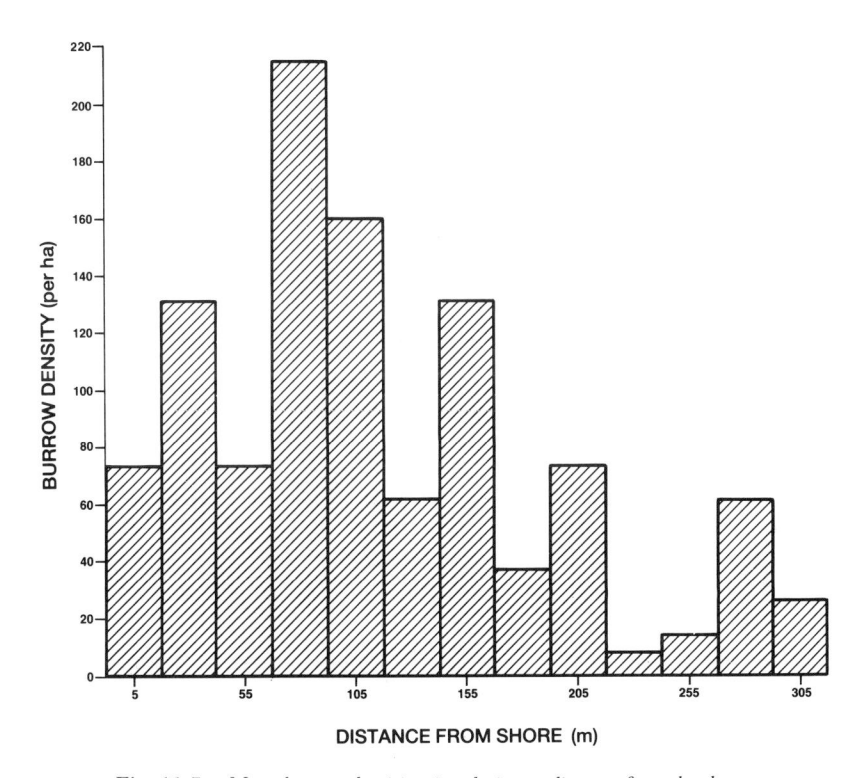

Fig. 11.7 *Mean burrow densities, in relation to distance from the shore.*

The marked effect of slope on burrow density is a little surprising, because many islands in the Queen Charlotte Islands on which Ancient Murrelets breed are much flatter than Reef Island (e.g. Rankine Island, House Island, and a number of islands in Skincuttle Inlet) and offer few areas with slopes of more than 30°. However, on some large islands, such as Ramsay and Lyell islands, the colony areas are particularly concentrated on steep slopes close to the sea (Rodway *et al.*, 1988), suggesting a preference for steep areas.

164 Studies at Reef Island

There are several reasons why steep slopes might provide better breeding habitat for Ancient Murrelets than flat areas. The ground is less likely to become waterlogged, passage through the canopy may be easier, allowing a more horizontal approach, and taking off from the surface may be easier, enabling the murrelets to avoid predators more easily. Presumably the flat islands where the murrelets nest provide other, compensating advantages.

Slope and distance from the sea are two factors that must contribute to the

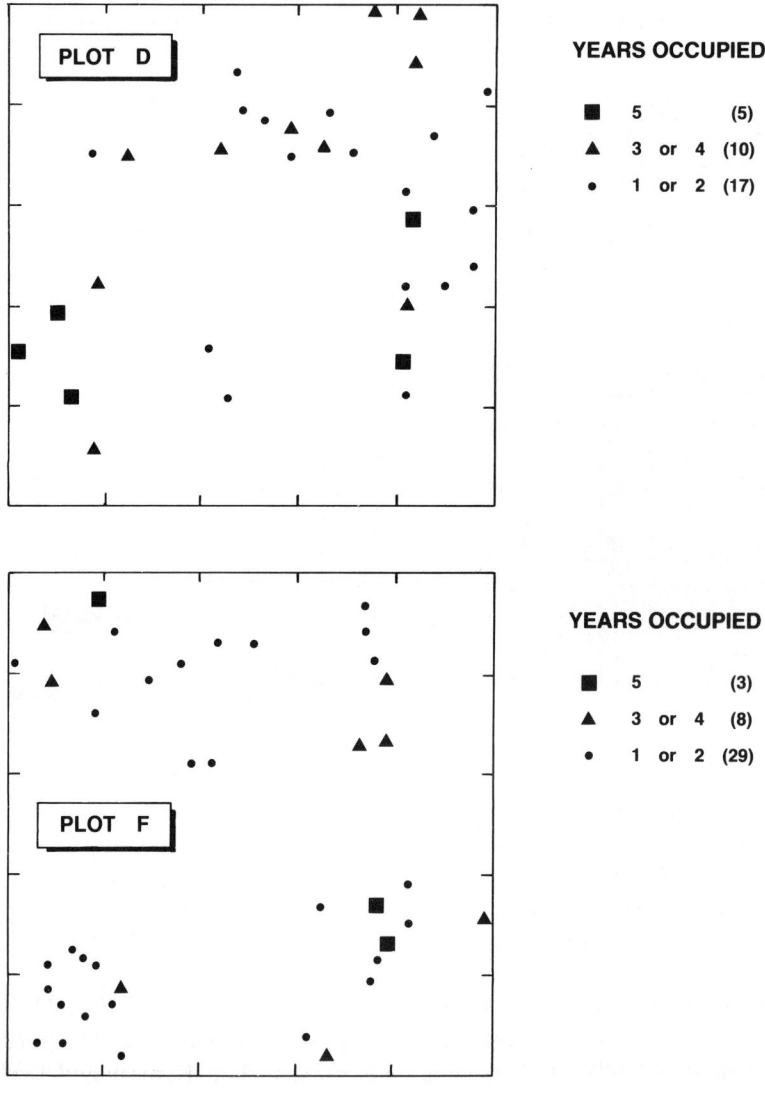

Fig. 11.8 *Frequency with which burrows in Plots D and F at Reef Island were occupied during 1985–89.*

Breeding habitat and burrows 165

observed clumping of burrows. However, clumping within the colony occurs at all scales. The number of burrows counted on whole transects in 1989 ranged from 0 to 18. Only 6 of 28 transects yielded 10 or more burrows, but these six accounted for 56% of all the burrows counted (Figure 11.5). On a smaller scale, we mapped the position of burrows on 5 m grids in several areas in 1984, and found that although 71% of 5 m squares contained no burrows, 13% contained two or more. The distribution of burrows occupied during the study at plots D and F, and the number of years of occupation, are shown in Figure 11.8, which illustrates the degree and scale of clumping in areas of relatively dense burrows.

Plot D, Reef Island; prime nesting habitat for Ancient Murrelets.

It was not hard to see why burrow density varied on the scale of transects spaced 50–100 m apart, because the steep slopes were cut by frequent ravines containing loose boulders and rubble, clearly unsuitable for Ancient Murrelet burrowing. Likewise, on the scale of 5 × 5 m squares, the position of suitable trees, stumps or logs clearly affected the distribution of burrows. However, on an intermediate scale, we also found differences between adjacent areas of similar topography and vegetation on the order of a few hundreds of square metres. In area D in 1984, a 25 × 25 m grid contained 28 burrows, a density of 448 burrows/ha, while in the nearby area E, actually steeper than D, a similar grid contained only 5 burrows (80/ha). Such uneven distribution of burrows was evident throughout the colony, and at this scale it may have been caused by social factors.

BURROW OCCUPANCY

When we began the study on Reef Island we put a lot of effort into discovering what proportion of burrows were occupied. In 1984 we examined four different areas, all with an above average density of burrows. We dug access tunnels to all burrows in an attempt to discover what proportion were being used. In one of the first areas that we chose (A), most proved to be unoccupied (Gaston et al., 1988), but the other three had occupancies ranging from 50 to 67%. This alerted us to the fact that occupancy rates could vary widely among different parts of the colony. Our later experience showed that A was not typical, but it demonstrated that certain areas can fall out of favour with prospecting murrelets. Our knock-down tags showed that unused burrows in plot A were visited almost as much as unused burrows in areas of much higher occupancy in 1984, but that the visitors apparently found the area unsatisfactory, because eggs were laid in very few (12% of 25). We never used knockdown tags on Plot A after 1984, but in 1989 only two burrows showed evidence of breeding, so presumably occupancy continued to be low.

We had additional evidence that not all parts of the colony were equally attractive to prospecting birds. Plot D, with a density of burrows similar to the colony average, had a normal occupancy rate in 1984 and 1985 but unoccupied burrows were entered only about one night in five. At Plot F, which had the same density of burrows as Plot D in 1985, but had a higher occupancy rate, we found that there was much more prospecting throughout the season, with unoccupied burrows being entered on average every third night (Figure 9.3). At plot F, new burrows were constructed throughout the period of the study, so that the total rose from 28 in 1985 to 47 in 1989 (a 68% increase). The number occupied in a given year remained stable, so that the proportion occupied declined. In contrast, at Plot D the number of burrows increased by only 31%, and the number occupied fell from 19 to 8, so occupancy declined precipitously.

At plot F we measured the lengths of occupied burrows in 1985 and again in 1989 (we laid a string from the mouth to the centre of the nest-cup and then measured the string). At seven of those occupied in both years the nest-cup was further from the entrance in 1989 than in 1985, compared to only two where it was closer; three were unchanged. Comparing the mean length of all occupied burrows in the two years, we found that in 1989 they averaged 18 cm longer than in 1985 (Table 11.2). Entrance diameters had increased slightly, but not significantly. At Plot A, most burrows seem to have been abandoned before we started work on Reef Island. It may be significant that burrows on Plot A were mostly much longer than the average on Plot D, which averaged longer than those on Plot F.

If many first-time breeders prefer to dig a new burrow, and the population remains stable, then a decrease in the proportion of burrows occupied in a given area seems inevitable. If birds sometimes extend existing burrows then

Table 11.2 *Ancient Murrelet burrow dimensions (cm), measured at Reef Island.*

Plot	Year	Entrance diam. Mean	S.D.	n	Burrow length Mean	S.D.	n	Chamber depth Mean	S.D.	n
D	1985	11.1	3.8	35	79.9	27.9	35	20.0	7.4	30
	1989	11.1	3.5	7	115.1	42.6	7			
F	1985	12.9	5.7	43	55.9	16.1	42	23.6	13.2	37
	1989	13.7	7.2	28	73.5	26.6	24			

old burrows are likely to be longer than new ones. Newly colonized areas should therefore be characterized by short burrows with a high rate of occupancy, while areas that have been used for many years will have long burrows, few of which will be occupied. Why Ancient Murrelets should dig new burrows when old ones are available is a mystery, and so is the process by which some birds begin to colonize an unoccupied area. Birds that have bred before nearly always return to the same burrow, or one close by (Chapter 14), so any movement in the area of the colony is dependent largely on the settlement of first time breeders.

CONTINUITY OF OCCUPATION

Although some new burrows are dug each year, the history of older burrows shows that they can be abandoned and reoccupied. Only 23% of the burrows occupied in 1984 ($n = 48$), and 27% of those occupied in 1985 ($n = 49$) at plots D and F were occupied continuously thereafter. Out of 67 burrows occupied in either 1984 or 1985, 34% were reoccupied after one or more years when they were not used.

In 14 cases the identity of both breeders in a particular burrow was known in more than one season. In the eight cases where the two seasons were consecutive, three pairs remained unchanged, four had changed one member, and in the other burrow both members of the pair had changed. Hence 10/16 (62%) were using the same burrow as the previous year. After two years one burrow contained the same pair, while two had changed one member. In each of the three cases separated by more than two years, one member had changed. Out of 58 birds trapped in burrows and retrapped in burrows in later years, 13 (22%) were in a different burrow. None of these birds relocated in newly excavated burrows; they all adopted ones that had been in existence during their first tenancy. In two cases pairs bred together in different burrows in different years, moving 2.5 and 10 m between sites. In one case two birds, previously paired, were found breeding separately in a later year, showing that divorce can occur. It may be more common than our observations suggest, because the chance of finding a bird again, if it left a study burrow, was fairly low. The relatively high rate of changes in pair membership indicates either that divorce

168   Studies at Reef Island

is common, or that mortality is high. Perhaps both are true. Where a burrow was left unoccupied for a year or more, we never found either of the original occupants when it became active again. It seems that once a murrelet abandons a burrow, it rarely returns to it.

Burrows where the pair had bred successfully were more likely to be occupied the following year than those where the eggs had been deserted (Figure 11.9). This was especially true of the 19 burrows where eggs were deserted in

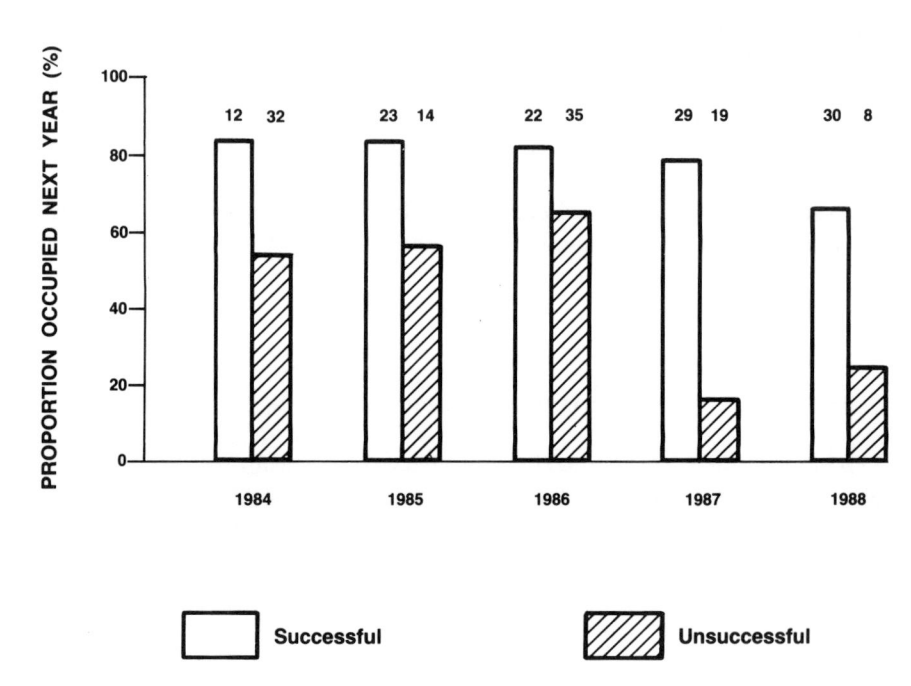

Fig. 11.9   *Proportion of burrows occupied in successive years, in relation to whether or not breeding was successful in the first year.*

1987, only 3 of which were occupied the following year. It is tempting to think that in 1987, when breeding was initiated very early, and all evidence pointed to a good breeding year (Chapter 13), many inexperienced birds tried to breed. If a large proportion of these did not attempt to breed the following year, when conditions seem to have been less favourable, or shifted their burrows, it would account for the sharp drop in the proportion of birds returning. The slight decline in the proportion of previously successful burrows reoccupied in 1988 and 1989, compared to earlier years, may be connected with the lower level of disturbance after 1986. This may have allowed more inexperienced birds to succeed than in earlier years. These birds were perhaps less likely to return the following year.

CONCLUSIONS

Ancient Murrelets on Reef Island place most of their burrows among the roots of trees, or stumps, or under fallen logs. Some of the burrows extend into crevices in the underlying rock. These features make them hard for a would-be predator to dig out. They take a number of precautions to conceal their sites and keep them clean and free from odour. This is the type of behaviour that we might expect to evolve in response to selection by predators hunting by sight and scent. The present dense and extensive colonies found in the Queen Charlotte Islands seem very vulnerable to ground predators, as the history of the Langara Island population suggests. However, it is possible that, in smaller numbers, and at lower densities, Ancient Murrelets might be capable of coexisting with some ground predators. Their persistence on Langara Island suggests that they can withstand considerable predation by rats. Small populations may also have survived to coexist with introduced foxes on some of the Aleutian Islands, especially by switching to nesting in rock crevices (D. Forsell, G.V. Byrd, pers. comm.).

The constant creation of new burrows, the tendency for new recruits to switch their preferred area of colonization, the tendency for birds to abandon burrows in which they were unsuccessful, and their reluctance to return to them later, could also help to minimize the effects of predation, although there are other plausible explanations for these phenomena. A difference in the likelihood that a breeding site will be reoccupied, in relation to the success of the breeding pair, is a fairly common observation for seabirds. The fact that Ancient Murrelets show relatively little sign of philopatry (Chapter 14) may help to explain the apparently rapid shift in the popularity of different areas for colonization. However, the subject of why Ancient Murrelets choose to nest where they do is still largely speculative.

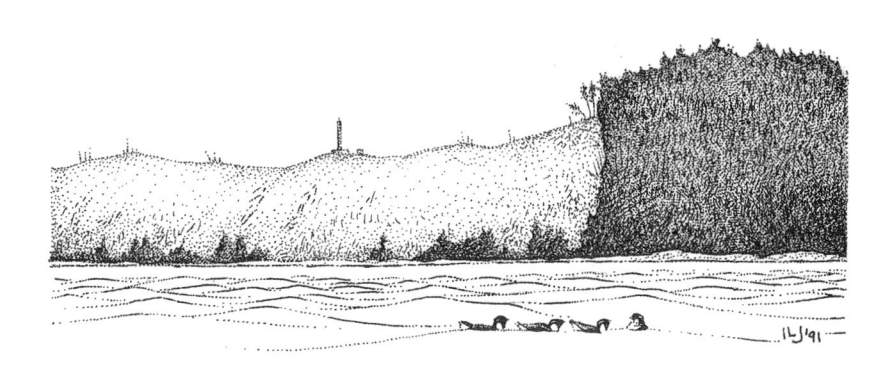

CHAPTER 12

Eggs and chicks

The eggs: size, colour and composition; clutch size and the interval between layings; egg recognition; incubation and hatching; the effect of chilling on hatching success and incubation period; time spent in the burrow by the chicks; some adaptations of the chicks.

THE EGGS; SIZE, COLOUR AND COMPOSITION

The eggs of the Ancient Murrelet, along with those of the other *Synthliboramphus* murrelets, are the largest eggs in relation to adult female body weight of any auk (Figure 12.1, see also Sealy, 1975a), or of any seabird except storm petrels (Croxall and Gaston, 1988). When we subtracted the weight of the average egg laid in a given year from the weights of females trapped with oviduct eggs in the same year, we found that the eggs constituted 22–23% of female weight at laying in four years (Table 12.1). This is identical with the proportion of egg weight to adult body weight found by Murray *et al.* (1983) for Xantus' Murrelet (22%).

Table 12.1 *Egg mass in relation to female body mass.*

Year	Female mass (g) Mean	S.D.	n	Egg mass (g) Mean	S.D.	n	Female minus egg	Egg as % of female mass
1984	244.5	14.0	30	43.8	3.2	57	200.7	21.8
1985	250.7	10.3	22	46.7	3.3	53	204.0	22.9
1988	247.4	7.1	16	45.3	2.6	41	202.1	22.4
1989	251.1	17.1	19	46.4	3.2	39	204.7	22.7

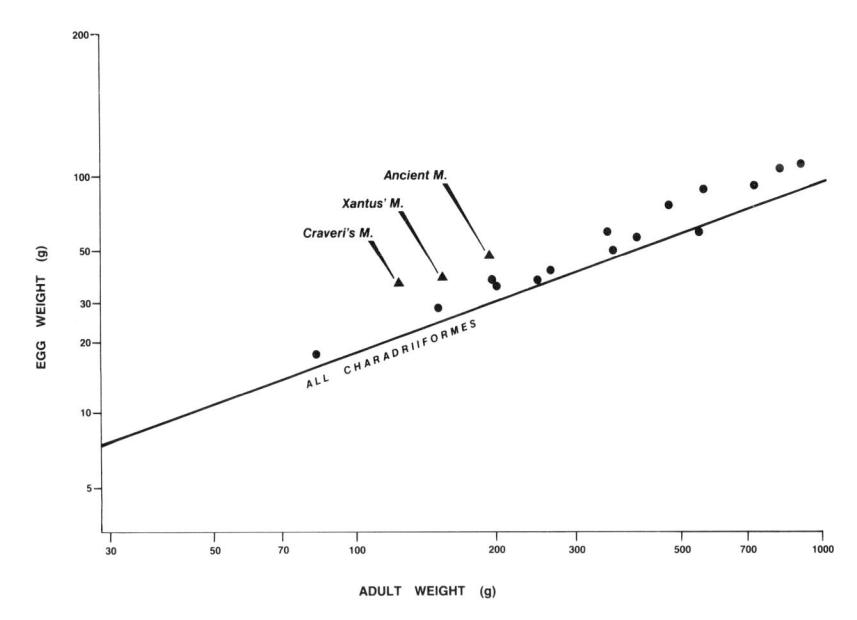

Fig. 12.1 *Fresh egg weight in relation to adult body weight in auks. Data points are from Croxall and Gaston (1988). The regression for all Charadriiformes is from Rahn et al. (1975).*

The mean rate at which female Ancient Murrelets form yolk during the 11 days that it takes for yolk deposition (3.07 g/day) is equivalent, in terms of energy stored, to 44% of the bird's standard metabolic rate (the amount of energy it expends in a day just to maintain itself), according to Astheimer and Grau (1990). This rate of yolk formation is higher relative to daily energy expenditure than they reported for 39 other species of seabirds, and more than double the rates for puffins and guillemots. This suggests that the process of egg formation must be a relatively taxing one for the female murrelet.

We found very little difference between the measurements and fresh weights of first and second eggs in the same clutch (Table 12.2). Consequently, in calculating mean egg weights and measurements for 1984–87, we used one egg from each clutch, taken at random with respect to order of laying. In later

Table 12.2 *Dimesions and mass of first and second laid eggs measured at Reef Island in 1984 and 1985. Only complete clutches of known order are included.*

Year	Order	Length (mm) Mean	s.d.	n	Width (mm) Mean	s.d.	n	Fresh mass (g) Mean	s.d.	n
1984	First	58.2	1.9	26	36.9	1.1	26	43.7	3.1	26
	Second	58.8	2.1	26	36.6	1.1	26	43.6	3.4	26
1985	First	59.2	2.0	24	37.9	1.0	24	46.2	3.1	24
	Second	59.8	1.9	24	38.0	1.0	24	47.0	3.0	24

years we only measured first eggs. Where the order of hatching was known, we found no tendency for the first laid egg to hatch first.

The eggs of the Ancient Murrelet are elliptical ovoid, with little difference in shape between the two ends. They differ from those of murres and razorbills which are distinctly pointed at one end. Those of other auks are generally less elongated. The fresh weight and measurements vary somewhat from year to

An Ancient Murrelet clutch, Reef Island.

year (Chapter 13) but show no sign of variation among the different colonies on the Queen Charlotte Islands. Mean measurements recorded elsewhere in the Queen Charlotte Islands, and also those reported from the U.S.S.R., all fell within the range of annual means recorded at Reef Island (Table 12.3). A small sample measured at Buldir Island by Vernon Byrd were distinctly larger (Table 12.3), but their date of laying was unknown.

The ground colour of the shells ranges from off-white to medium brown, sometimes bluish, but most are some shade of buff or pale brown. Markings, usually irregular spots and blotches 1–5 mm across, are mostly earth-brown and scattered more or less randomly. On some eggs they form a diffuse ring at one end. In appearance, the eggs are most similar to those of the guillemots, *Cepphus* spp, both in shape and colour. The surface texture is remarkably smooth, feeling almost slippery to the touch. This can be annoying when you are trying to extract an egg from a burrow at the end of your reach, with the tips of your fingers. It may contribute to the difficulty that deer mice have in breaking into them (Chapter 14).

Table 12.3 *Egg size in Ancient Murrelets.*

Area	Year	Fresh mass (g) Mean	S.D.	n	Length (mm) Mean	S.D.	n	Breadth (mm) Mean	S.D.	n	Reference
QUEEN CHARLOTTE ISLANDS											
Reef I.	1984	43.8	3.2	57	58.7	2.0	98	37.1	1.2	98	
	1985	46.7	3.3	53	59.3	1.9	99	37.9	1.1	99	
	1986	45.7	2.8	52	59.3	2.1	52	37.6	1.1	52	This study
	1987	47.6	3.4	50	60.0	2.2	50	38.1	1.1	50	
	1988	45.3	2.6	41	59.3	2.0	41	37.5	0.9	41	
	1989	46.4	3.2	39	59.3	2.4	39	37.9	1.5	39	
Ramsay I.	1984				59.4	1.7	39	37.3	1.1	39	Lemon and Rodway (pers. comm.)
Frederick I.	1981				60.3	2.9	53	37.8	1.5	53	Vermeer and Lemon (1986)
Langara I.	1970	44.9	—	15	59.4		200	37.4		200	Sealy (1976)
	1988				58.4	1.8	13	37.9	1.2	13	Bertram (1989)
Helgesen I.	1986	44.3	3.8	10	58.2	2.2	22	37.5	0.9	22	Rodway *et al.* (1990)
ELSEWHERE											
Buldir I., Alaska	1974–76				62.2	1.4	8	39.0	0.7	8	G.V.Byrd (pers. comm.)
U.S.S.R.	?				60.0		72	36.5		72	Flint and Golovkin (1990)

As well as being relatively large, the eggs of the Ancient Murrelet differ from those of most other auks in their composition. To examine this, we removed a number of eggs that had been deserted before incubation began, hard-boiled them and then separated shell, yolk and albumen and weighed them individually. These components made up 8, 45 and 47%, respectively, of the fresh weight, giving a value of 1.03 for the ratio of yolk to albumen. This compares with ratios of 0.52–0.66 among other auks that have been examined (Birkhead and Gaston, 1988). Both the relatively large size of the egg and the high proportion of yolk in the contents are characteristic of precocial birds, when compared with those with altricial young (Carey *et al.*, 1980).

CLUTCH SIZE AND THE INTERVAL BETWEEN LAYINGS

The normal clutch is two. Some accounts in the literature mention one or two, but incubated clutches seem to be nearly always of two eggs; we saw only 8 single-egg clutches out of 298 (3%) on Reef Island. Four other clutches were of 3 or 4 eggs, but the ground colour and markings indicated in all cases that these were the product of two females. Most clutches recorded by Sealy (1976;

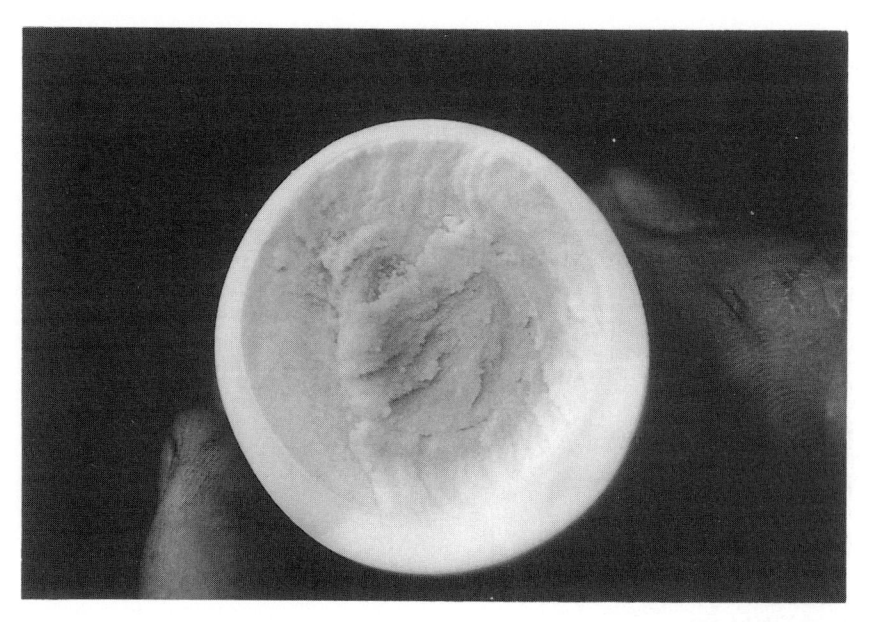

Section through the middle of a hard boiled Ancient Murrelet's egg, showing the very large size of the yolk.

A clutch of eggs probably resulting from the laying of two females.

95% of 154), and by Vermeer and Lemon (1986; 95% of 192) were also of two eggs, as were 42 clutches found by Vyatkin (pers. comm. via N. Litvinenko). We found a few eggs laid on the surface every year, and it seems possible that these may have been laid by females that were unable to find their burrow, hence accounting for some at least of the one-egg clutches. We had no evidence that replacement clutches were ever laid, and nor did Sealy. Bent (1919), based on the observations of A.C. Howell, considered that Xantus' Murrelets were capable of rearing two broods in a season. However, Murray et al. (1983) never found pairs rearing more than one brood in a season, although they found definite evidence of occasional replacement laying.

According to Sealy, the eggs are laid about one week apart (average of 18 clutches, 7.2, range 6–8 days). We found that the average interval at Reef Island was 7.8 days ($n = 11$) in 1984 and 8.2 days ($n = 12$) in 1985, with a range of 6–10 days. We did not measure the interval in other years. In 1989 the time elapsed between the laying of the first egg and the start of incubation averaged 8.7 days (excluding one period of 3 days, see below). If the period of neglect before the start of incubation was similar to that in 1984 and 1985 this suggests a laying interval of 7–8 days. I have assumed an average of 8 days in extrapolations of laying dates. Murray et al. (1983) found a similar 8 day interval (range 5–12 days) between the laying of the two eggs of Xantus' Murrelet in California.

Incubation of one-egg clutches at Reef Island began 7–13 days after the egg was laid. Incubation of two-egg clutches started an average of 0.5 days after clutch completion in 1984, and 1.6 days in 1985, but at about a third of nests incubation began immediately after the second egg had been laid. At the rest, the time elapsed ranged up to 13 days.

EGG RECOGNITION

In 1988 David Powell carried out some tests to see whether the appearance of the second egg caused the male to begin incubating. He stained hens' eggs, by boiling them in a saucepan with old tea-bags, to make them similar to Ancient Murrelets' in ground colour, and added them to nests immediately after the laying of the first egg. We checked the nests daily thereafter to see whether incubation had begun, terminating the experiment by removing the hen's egg after six days.

We never induced incubation to begin; instead two of the foreign eggs were shifted out of the nest-cup, one being kicked right outside the burrow. This led us to believe that one of the parents had identified it correctly as foreign. A similar case occurred in 1985. Three eggs were laid in one nest, one being clearly dissimilar in colour from the other two. When found, the odd egg was lying in the nest-chamber but outside the cup. I placed it in the cup, hoping to see whether the birds could incubate all three, but at the next inspection it was again displaced from the nest. I placed it in the nest-cup once more, and it was again displaced, suggesting that the birds had deliberately pushed it out. At that point I gave up, thinking it was a chance occurrence. The rejection of the chicken eggs made me think again. It seems possible that Ancient Murrelets are capable of distinguishing their own from other eggs under certain circumstances, although how they do it, in the darkness of the nest chamber, remains a mystery.

INCUBATION AND HATCHING

Like all birds' eggs, those of the Ancient Murrelet lose weight by the evaporation of water throughout the incubation period. As the volume remains constant, the density of the egg decreases, giving rise to the flotation method of determining how long an egg has been incubated (Nol and Blokpoel, 1983). The decrease in density is practically linear, allowing it to be used to predict the approximate date of the start of incubation (Figure 12.2). We used this technique (outlined by Collins and Gaston, 1987) to estimate approximate dates of laying for samples where we measured and weighed some eggs after the start of incubation, as happened at Reef Island in 1986 and 1987.

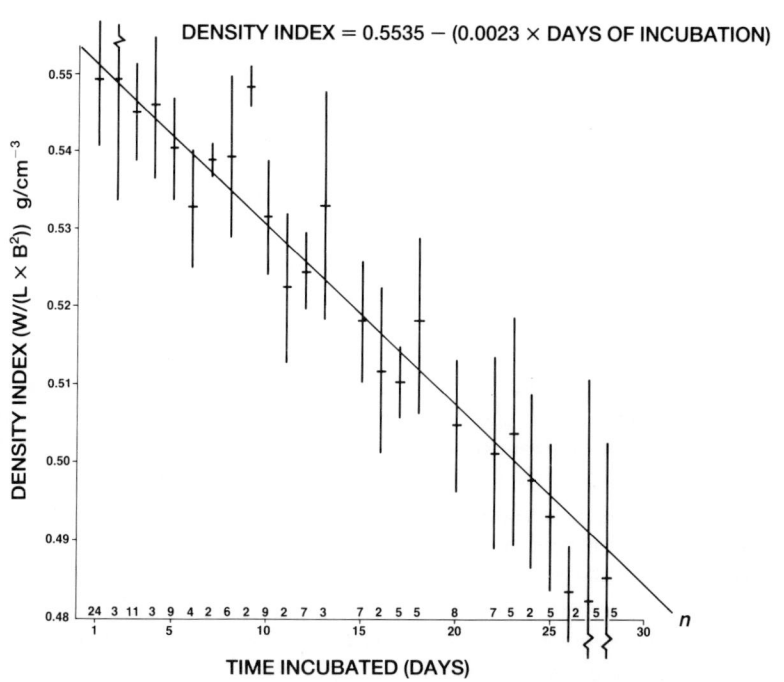

Fig. 12.2 *Egg density indices in relation to the length of time for which the egg had been incubated.*

The incubation period is generally measured from the completion of the clutch to the hatching of the first egg. Because we did not inspect burrows after the laying of the first egg in later years at Reef Island we could not measure actual incubation periods. Instead we were able to measure the number of days of incubation that the eggs received, which is usually somewhat less in these birds. At Reef Island most Ancient Murrelet clutches of one and two eggs hatched after 29–31 days of incubation (Table 12.4). At one nest with three eggs in 1989, where incubation continued to term (the other clutches of more than two eggs were either deserted immediately or incubated only intermittently before being deserted), one egg hatched after 36 days of incubation, and the parents departed with the chick two days later. Presumably the extra days of incubation were necessary because all three eggs were being rotated. The breeding pair were obviously uneasy about the enlarged clutch, because the eggs at this nest were neglected for a total of 10 days after incubation had begun; 6 days more than at any other nest in that year.

Table 12.4 *Incubation periods and number of days for which eggs were actually incubated.*

Year	Incubation period										Days incubated					
	29	30	31	32	33	34	35	36	37	28	29	30	31	32	33
1984		1	2	1			2			1	2	2	1		
1985	1	1	3	2	1	2	1	2	1	1	1	6	5		1
1988											4	15	7		
1989											10	44	9	3	
Totals	1	2	5	3	1	2	3	2	1	2	17	67	22	3	1

Incubation periods that we observed in 1984 and 1985 ranged from 29–37 days. Sealy observed 33–47 days ($n = 34$) at Langara Island, but 74% of periods were 33 or 34 days. He found that the chicks hatched within an hour of one another at all but one of 62 nests; at the remaining burrow they hatched about 6 h apart. As we checked burrows only every 24 h, we could not judge hatching synchrony as accurately. In most instances the two chicks were both hatched when first found, and hence must have hatched within 24 h of one another. However, in 5 cases out of 26 in 1988 and 6 out of 35 in 1989, we found one chick hatched, while the other egg was still pipping. In all but one case, the eggs had hatched by the next day. Hence the difference could have been anywhere from a few minutes to almost 48 h. In the one other instance, two days elapsed before the second egg was found hatched. In this case the difference must have exceeded 24 h. In 8 out of the 11 cases where we found one chick hatched and the other not, the chick's down had dried, suggesting that it had been out of the egg for a minimum of several hours. Our results suggest that the hatching of chicks at Reef Island was less synchronous than those observed by Sealy.

178   Studies at Reef Island

At one other nest at Reef Island, incubation began only three days after the laying of the first egg. When the first egg hatched, the second was beginning to crack, but it was still pipped the next day; the day after that the parent and chick had gone, and the still unhatched egg was found partially eaten by mice. This is a vivid illustration of the importance for the murrelets of not commencing incubation until the clutch is complete.

The difference between the length of the incubation period and the number of days of actual incubation was caused by the murrelets' habit of neglecting their eggs for a few days, usually immediately after the completion of the clutch. Natural neglect after the beginning of incubation could only be measured in 1988 and 1989, because of the possibility that our activities induced some of the neglect observed in earlier years. In 1988 54% ($n = 26$) and in 1989 57% ($n = 35$) of clutches were neglected for at least one day after the start of incubation, but in 68% of cases the neglect occurred after only one or two days of incubation. The total number of days for which clutches were neglected ranged from 1 to 10 days, but only 21% were neglected for more than two days (Figure 12.3).

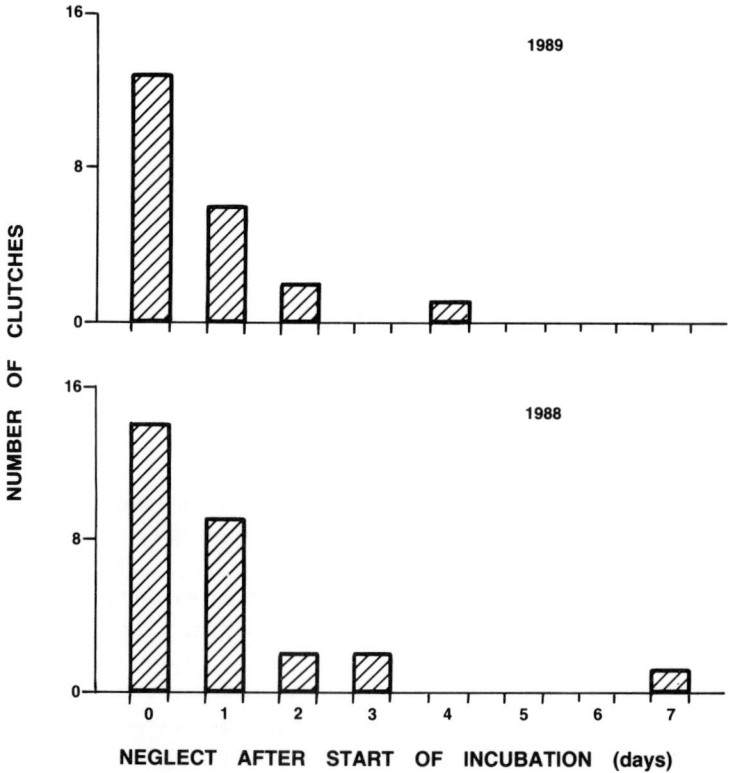

Fig. 12.3   *Numbers of days of neglect observed after the start of incubation at burrows checked only with temperature probes (hence undisturbed).*

THE EFFECTS OF CHILLING ON HATCHING AND INCUBATION

In 1988 and 1989 we observed less neglect than in 1984 and 1985, particularly after the clutch had been incubated for several days. During periods of neglect the temperature of the eggs fell to the ambient temperature of the burrow, usually 6–10°C. Despite this chilling, nearly all eggs that were not deserted completely eventually hatched. We concluded that Ancient Murrelet eggs were very resistant to chilling, and in 1988 we designed an experiment to demonstrate just how resistant they were (Gaston and Powell, 1989).

At a sample of burrows we removed one egg after 10 days and placed it in a nearby unoccupied burrow, so that the eggs remained at the ambient temperature and humidity that they might experience during a period of neglect. The eggs were returned to their burrows after 48 h, and the other egg in each clutch was then removed for a similar period. We performed the same treatment after 20 days at another sample of nests. Thirteen out of 14 eggs temporarily removed, including all six removed at 20 days, hatched successfully. The egg which did not hatch showed no sign of development when examined and hence must have been infertile. For comparison, we performed a similar test with Thick-Billed Murres breeding at Coats Island, Northwest Territories, where ambient temperatures during incubation are very similar to those recorded at Reef Island. In this species, eggs temporarily removed after five days of incubation hatched successfully, but none of those removed after more than 10 days of incubation did so, showing that Ancient Murrelet embryos have a greater resistance to chilling than those of Thick-billed Murres. In fact, among seabirds for which we have information, petrels are the only ones with embryos which can match those of Ancient Murrelets for resistance to chilling (Boersma and Wheelwright, 1979).

Interestingly, at the three Ancient Murrelet nests where eggs were removed after 20 days, the clutches all required 32 days of incubation before hatching. For 30 other clutches monitored that year the number of days of incubation did not exceed 31 (Gaston and Powell, 1989). This result suggests that Ancient Murrelet eggs neglected late in the incubation period suffer a penalty of some kind, causing their development towards hatching to slow down. This contrasts with the situation found by Murray et al. (1983) for Xantus' Murrelet, where the total number of days of incubation fell somewhat with increasing periods of neglect. The mean ambient temperature at their study site during the incubation period exceeded 20°C, and this may have been sufficient to allow some development to proceed without parental incubation.

A feature of Ancient Murrelet biology which may be related to the great resistance of their eggs to chilling, is the relatively small size of the brood patches. The maximum diameter of the patch, at about 25 mm, is only half the length of an egg. Hence no more than half of one surface of the egg can be in contact with the brood-patch at any one time. All of the auklets, except the Least, which is a much smaller bird, have larger brood-patches than the

Ancient Murrelet, although they lay smaller eggs (Manuwal, 1972). The murres, which lay larger eggs, have brood-patches more than twice as large as Ancient Murrelets (pers. obs.), so that more than 75% of the length of the egg can be in contact with the brood patch. It appears that Ancient Murrelet eggs can develop with a much more localized input of incubation heat than other auks. They may alter the position of their eggs more frequently, but we have no information on this.

TIME SPENT IN THE BURROW BY THE CHICKS

As for hatching, the length of time spent in the burrow by the chicks could not be measured exactly because we inspected burrows only once a day. Sealy reported that most hatching occurred at night. However, we found a few chicks still wet at inspections made in mid-afternoon, so we know that some hatched in the middle of the day at Reef Island. As the nights are very short in late May at the latitude of the Queen Charlotte Islands, a second chick, failing to hatch on the same night as its sibling, would have to wait almost 24 h before hatching, if it confined itself to doing so at night. This does not appear to be a very advantageous strategy, and it seems unlikely that hatching is as closely tied to the hours of darkness as Sealy believed.

Most of our burrow inspections were made between 12.00 and 17.00 h, and most chicks departed between midnight and 02.00 h on the second night after they were found. The minimum period spent in the burrow by these chicks was in the region of 36 h, and the maximum about 60 h. Assuming that most hatched during the night, the departures would have been about 48 h after hatching. In all years for which we kept records, a few chicks left on the first night after they were found, and a few remained more than two days (Figure 12.4). Many of these were members of broods where both chicks were not found hatched on the same day. Both members of the brood invariably left the burrow together.

SOME ADAPTATIONS OF THE CHICKS

As is the case for their eggs, the young chicks of the Ancient Murrelet show some unusual characteristics associated with their precocial habit. Coming from a big egg, the newly hatched chick is also rather large; the mean weights of chicks when first found ranged from 30.3 to 32.7 g over the six years of the study, which is about 15% of adult body weight, about 67% of fresh egg weight. None of the other auks for which information is available have chicks that exceed 11% of adult body weight at hatching (from Duncan and Gaston, 1988).

The chick is covered with a dense coat of down, strikingly more closely

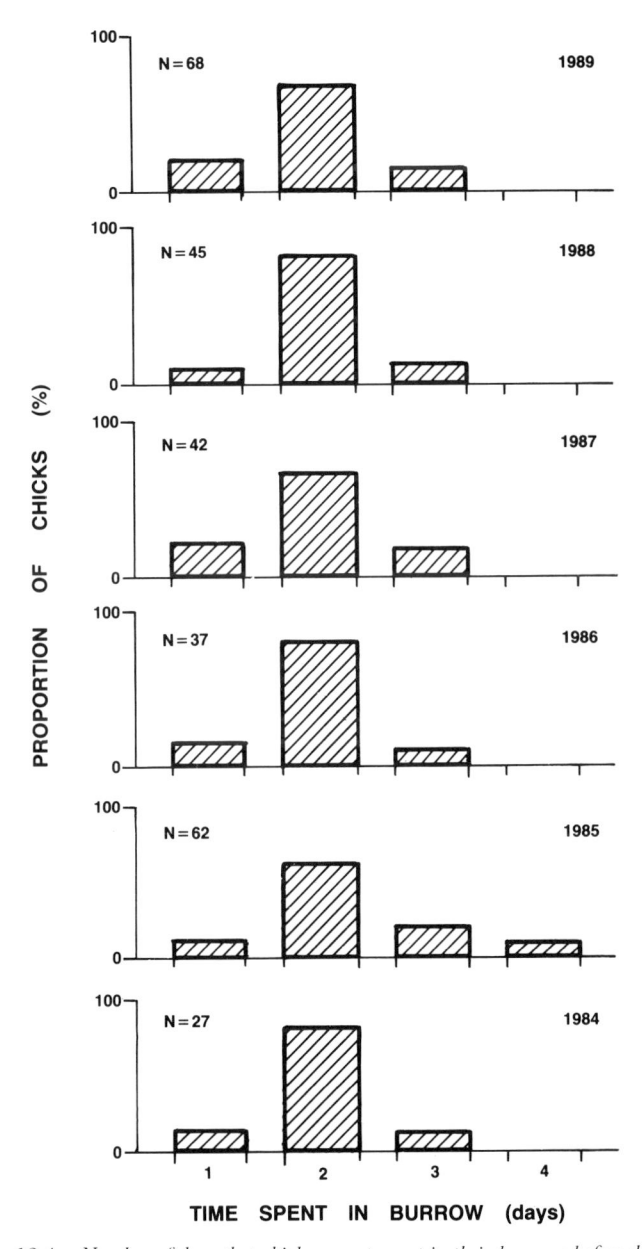

Fig. 12.4 *Number of days that chicks were present in their burrows before departure.*

packed than that of other auk chicks that I have handled, especially on the underparts. In colour, the down resembles the winter plumage of the adults, being black on the head, shading to grey on the back, and white underneath. The bill is dark horn with a prominent pale egg tooth. The feet resemble

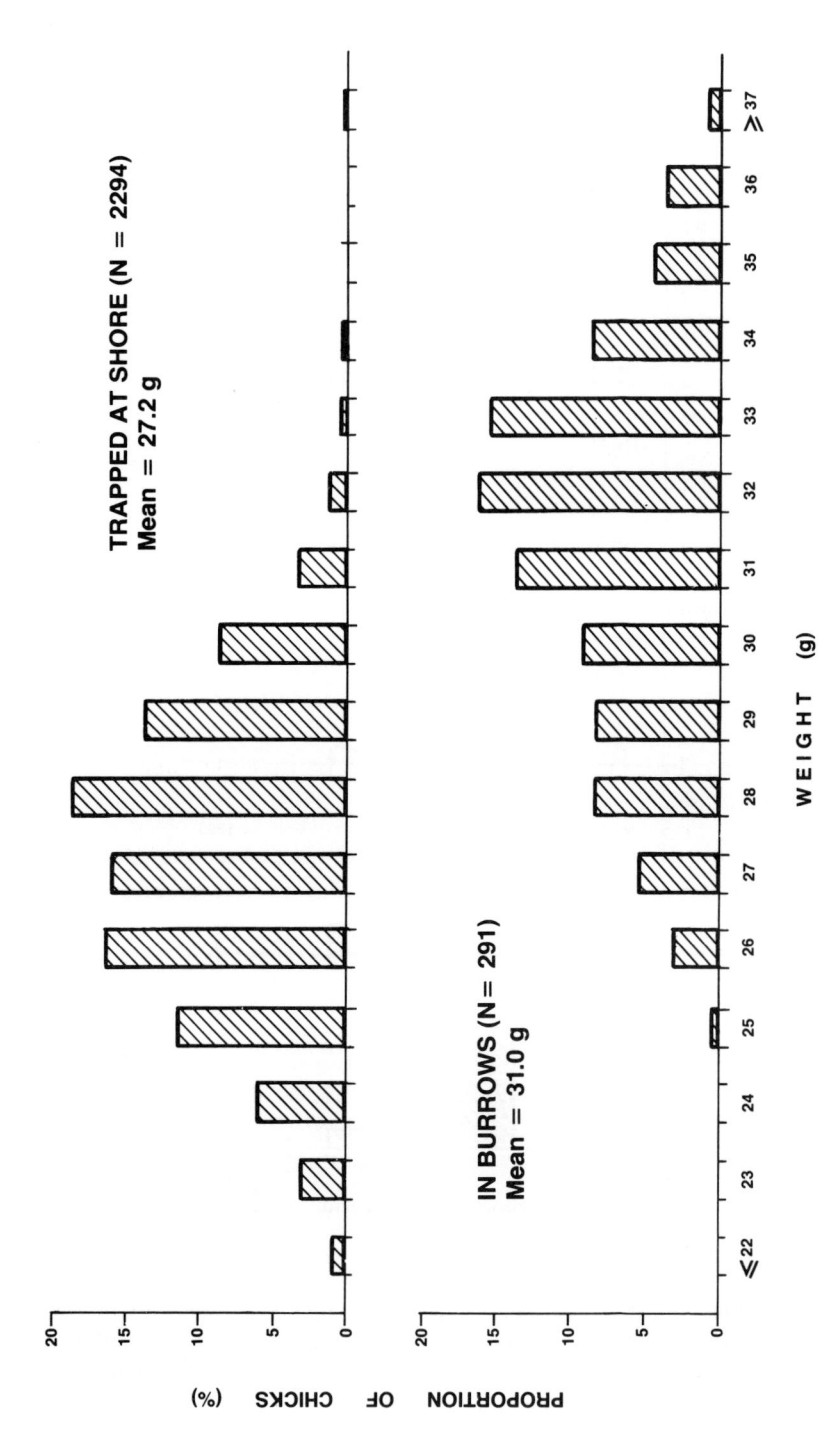

Fig. 12.5 *Weights of chicks removed from burrows on the first day after hatching, compared with the weights of chicks trapped at the shore during departure from the colony.*

those of the adults very closely, being similarly bluish flesh coloured. As Sealy has pointed out, the feet and legs are virtually adult size at hatching, the tarsus being 93% of adult length. The total mass of the legs at this stage is 27% of the carcass (i.e. the plucked chick minus the yolk sac; Duncan and Gaston, 1988).

To fuel the long journey that they have to undertake before receiving their first meal, Ancient Murrelet chicks at hatching have a very large store of fat, which comprises 42% of their dry weight after the feathers have been removed (Duncan and Gaston, 1988). About a quarter of this fat remains in the yolk sac, from which all of it originally derived and which comprises about 10% of its fresh weight. The rest, apart from what is incorporated into muscles and other tissues, is distributed about the body, forming a fairly even subcutaneous layer similar to that which can be seen under the skin of most adult auks. As well as functioning as an energy reserve, this layer of fat may also help to insulate the chick against the cold. The chicks of Xantus' Murrelet have been shown to be exceptionally good at regulating their body temperature in cold water (Eppley, 1984), and Ancient Murrelets are probably even better, since they generally inhabit colder latitudes. An analysis of four other species of auks showed that the fat content of their chicks at hatching was lower than that of the Ancient Murrelet, comprising from 22 to 33% of their dry weight. In the case of the Thick-billed Murre, the relative size of the yolk sac was larger than that of the Ancient Murrelet, comprising 17% of fresh weight, and containing 54% of the chicks lipid reserve. Hence, although the Ancient Murrelet chick carries a lot of fat at hatching, its yolk sac is not exceptionally large, most of the contents having been redistributed elsewhere.

Chicks weighed in burrows on the first day after hatching averaged 31.0 g in weight (range 25–37 g). Repeat weighings of a few chicks showed that they lost weight at about 2 g/day while they remained in the burrow. As most chicks probably hatched about 12 h before their first weighing, we can assume that the average weight at hatching was about 32 g. By the time that they reached the shore, average weights had fallen to 27.2 g (range 19–38, Figure 12.5). We compared the weight losses of chicks reweighed in their burrow with those of chicks weighed in the burrow and retrapped at the shore, en route to the sea. We found that the rate of loss in weight of those retrapped at the shore was higher than the rate of loss in the burrow (Figure 12.6; the dot–dash line shows the probable change in weight between hatching and departure), suggesting that loss of weight during the last 10 h before departure is higher than during the preceding period. Increased weight loss may be associated with an increase in activity and a reduction in the amount of time spent being brooded by the parent at this stage (Chapter 8). Ancient Murrelet chicks contain an average of about 3 g of fat at hatching (Duncan and Gaston, 1988), but they lose 40% of this before leaving the burrow, so only 1.8 g remains by the time they reach the sea (Duncan and Gaston, 1990). About 1.2 g is metabolized

184   Studies at Reef Island

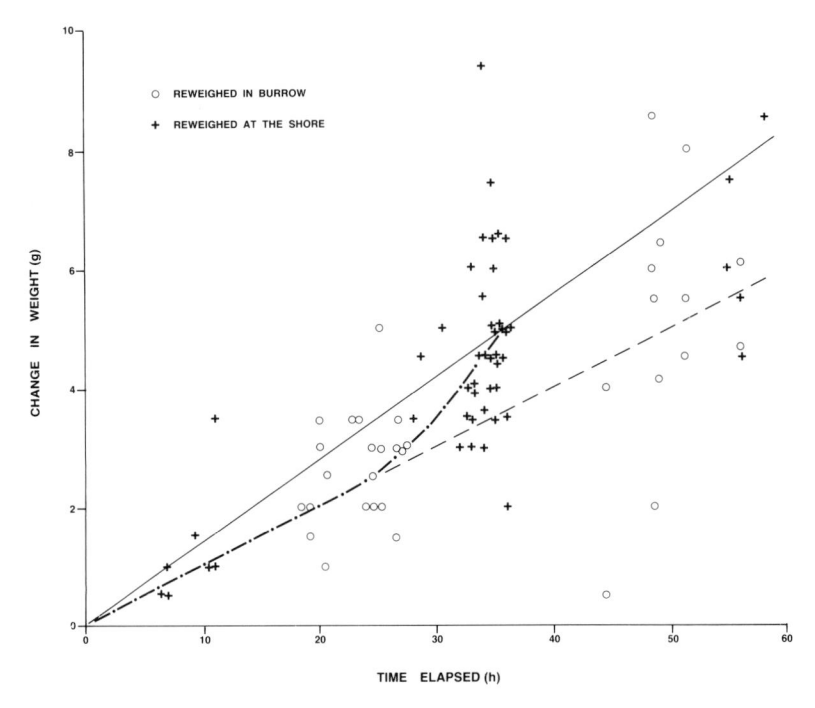

Fig. 12.6 *Weight losses of chicks reweighed at different intervals after the first day in the burrow. The dashed line shows the regression for chicks retrapped in the burrow, and the solid line for chicks retrapped at the shore. The curve, fitted by eye, shows the likely relationship for an individual chick.*

during the period between hatching and colony departure. The rest of their weight loss is presumably due to a reduction in the amount of water in the tissues, a process that occurs as the chicks mature (Ricklefs, 1979).

The remaining reserve of fat is important in enabling the chick to undertake the long, steady swim that will take it to the feeding area. We do not know exactly how much energy the chicks use in swimming, but a rough estimate by David Duncan suggested that an average chick uses up most of its strategic reserve in completing the 18 h swim from the colony (Duncan and Gaston, 1990). We must assume from this that those chicks which remain longer than usual in the burrow, and therefore use up a greater proportion of their fat reserves before departure, may be incapable of completing the long journey to the feeding area. This presumably explains why chicks that are abandoned by their parents will still leave the burrow after about two days (Chapter 8). Once they are ready to leave, the longer they remain in the burrow, the less chance they have of making a successful departure.

CHAPTER 13

Timing of breeding and its effects

How timing of breeding was recorded; the effect of early vs late years on egg size, and adult and chick weights; possible links to other events in Hecate Strait; correspondence between the laying dates and egg volumes of individual birds in different years; seasonal changes in chick weights at departure

Most people accept that natural selection has moulded the reproductive behaviour of birds so that they lay their eggs at the time of year when they are likely to produce the maximum number of surviving offspring (Lack, 1968). In order to do this, the birds use clues from climate, day-length and the condition of the environment so that they can initiate breeding at the appropriate moment. Generally, day-length, affecting hormonal cycles, is the dominant factor determining when the birds enter breeding condition (Murton and Westwood, 1977). Environment, operating through the nutritional condition of the birds, usually causes small-scale adjustments within the broad pattern set by inherited responses to day-length. In field work, it is the residual variation which is caused by day-to-day and year-to-year fluctuations in weather and feeding conditions that is of most interest, as it may tell us something about how the birds respond to environmental changes.

We recorded the timing of breeding at Reef Island in several different ways. In years when we were present from the start of of the season, we examined our study burrows daily to record the laying of the first egg. Once we began to use temperature probes to monitor incubation, we could not observe the exact date of clutch completion, and instead we had to estimate it from the date of the onset of incubation. Our information on the number of days that eggs were neglected during incubation, and the number of days spent by chicks in the burrow before departure, also allowed us to estimate dates of laying by working backwards from the date that chicks were trapped in our funnels. To obtain the estimated date of clutch completion, we subtracted 34 days from the date of departure—two days of neglect, 30 days of incubation, and two days in the burrow. To obtain the date at which the first egg was laid, we subtracted a further eight days. Although the accuracy of these extrapolations varies somewhat for individual burrows, it yielded a mean value identical to that obtained by direct inspection in 1987–89, when we had comparable data.

TIMING AND SPREAD OF LAYING

The earliest date of laying at Reef Island was 30 March, and the latest 21 May. Both these extreme records were of females caught arriving with eggs ready to lay. The extreme dates observed in our study burrows were 1 April and 16 May. The earliest that chicks were recorded leaving the island was 8 May, in 1987, which suggests a date of first egg-laying of 28 March. We stopped trapping chicks at the end of the season after we had had a night without any departures, so a few chicks probably left after we had ceased operations. The latest chick we actually trapped was on 16 June, in 1986, suggesting that the second egg of that clutch was laid about 13 May.

During the four years when we trapped chicks in our funnels throughout the departure period, 50% of chicks left over periods of from 6–10 days, and 90% over 15–25 days. In 1987, the earliest year, departures were particularly synchronous, with nearly 30% of the chicks leaving in just three nights (Figure 13.1). In 1988 and 1989 there were two peaks of departures, separated by small troughs, which coincided with the median dates of departure.

To look for differences in timing of breeding between different parts of the colony, we compared median dates of laying at plots D and F, but found no consistent difference. Moreover, the median dates of capture of chicks in the different funnels were identical in 1987 and 1988, and within one day in the others (Table 13.1). It appears that whatever factors affect the timing of laying of Ancient Murrelets at Reef Island, they operate uniformly in different parts of the colony. This contrasts somewhat with what has been observed in some other auks, where the timing of laying may vary consistently between local groups within the same colony (Birkhead and Harris, 1985).

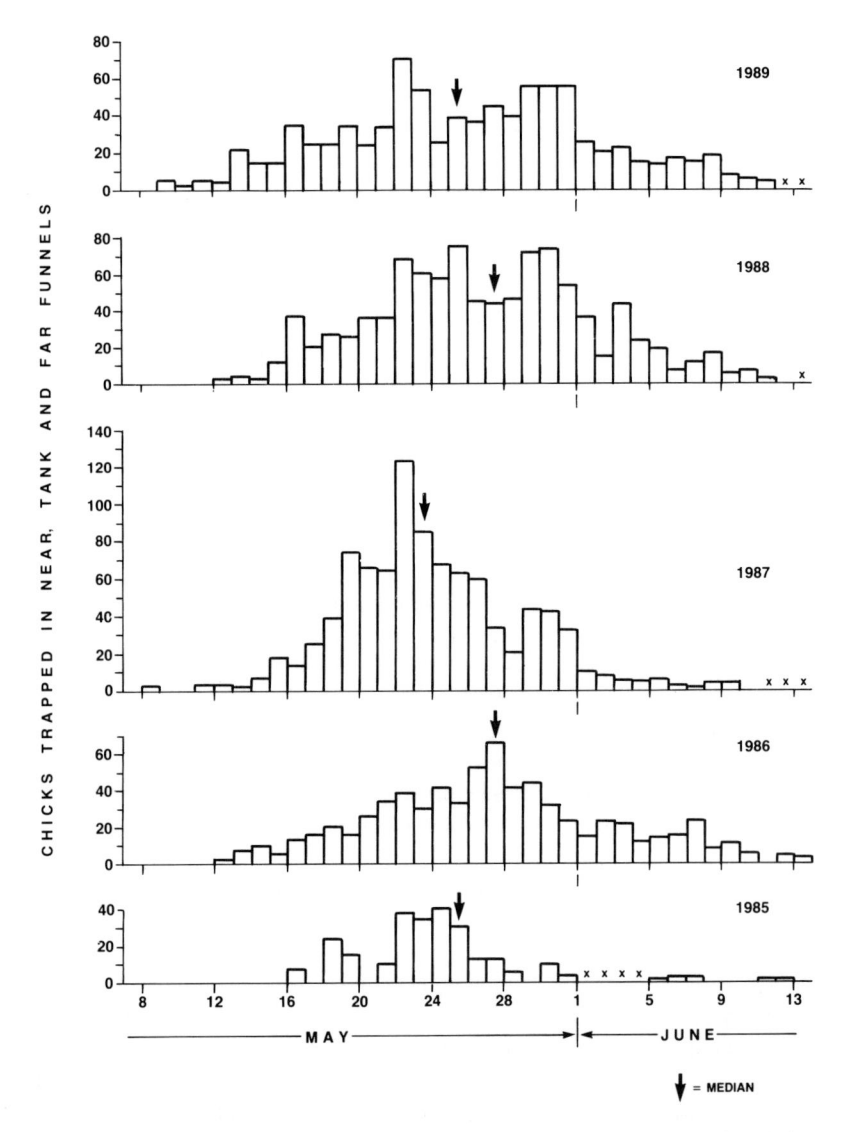

Fig. 13.1 Dates of chick departures at Reef Island, based on captures in trapping funnels (x = no trapping).

Table 13.1 *Median dates of departure of chicks captured in three different funnels (numbers captured in parentheses).*

	Median departure date (May)		
Year	Near	Tank	Far
1986	27	26	27
	(185)	(151)	(376)
1987	23	23	23
	(226)	(189)	(524)
1988	27	27	27
	(207)	(255)	(525)
1989	27	26	25
	(182)	(240)	(473)

YEAR-TO-YEAR VARIATION

With only six years of observations, we did not obtain much information on inter-year variation. The earliest year was 1987, with a median date of clutch completion of 19 April. The latest was 1984, when the median was just nine days later. Despite this rather modest variation in the timing of laying, we found that it correlated surprisingly well with other aspects of the birds' breeding biology, indicating that it reflected important year-to-year differences in environmental conditions.

We found that egg size was closely associated with the timing of laying (Figure 13.2). The mean fresh weight was highest in the earliest year, 1987 (47.6 ± 3.4 g), and lowest in the latest, 1984 (43.8 ± 3.2 g). The regression of egg weight on median laying date predicted a 0.4 g drop in fresh egg weight for every day that the median date of laying was retarded. In every year there was a slight tendency for eggs laid later in the season to be smaller than those laid earlier, but this effect was much smaller than the inter-year effect, the regressions predicting only a 0.1 g decrease in fresh egg weight for every day later that they were laid (Figure 13.3). The weights of females trapped with eggs ready to lay also showed a decline with date, significant in 1989 ($r = -0.62$, 16 df, $P < 0.01$).

If differences in the ease with which birds could obtain food were responsible for year-to-year differences in median laying dates, then the observed correlation between median date of laying and median egg size makes sense. Presumably, in years when food is less readily available, females delay laying because it takes them longer to build up the necessary reserves of nutrients. They may compromise by laying slightly smaller eggs slightly sooner than they would otherwise have been able to, a trade-off which has been described for other auks (e.g. Thick-billed Murre; Birkhead and Nettleship, 1982).

We might also expect differences in the availability of food to be reflected in adult body weights, and this turns out to be the case. Adult weight increases in

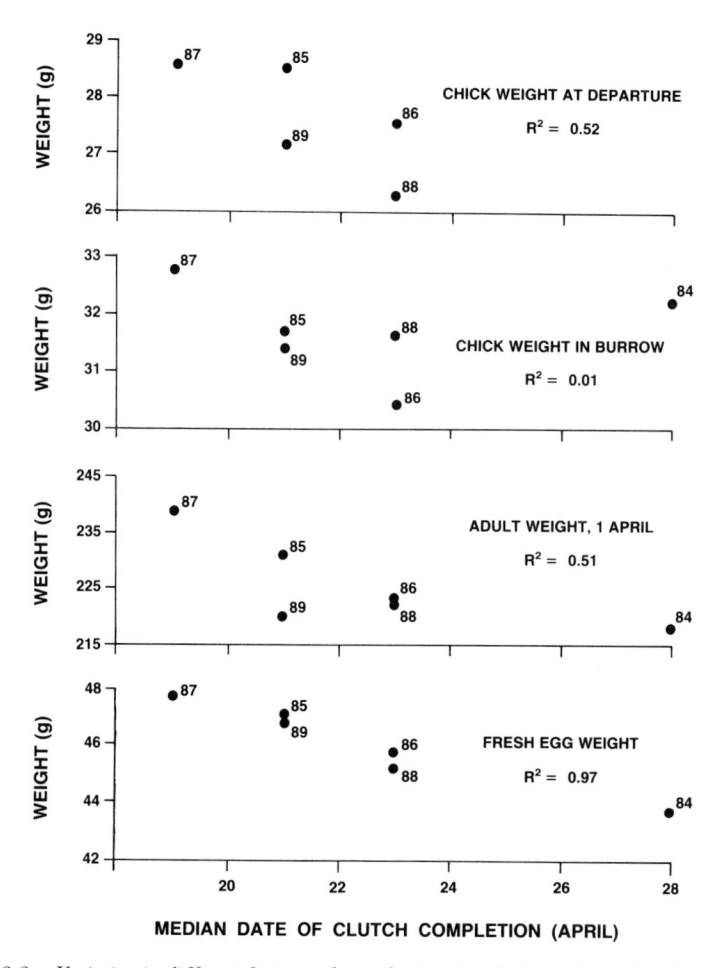

Fig. 13.2 *Variation in different features of reproduction, in relation to the median date of clutch completion.*

the pre-laying period and then declines during incubation (Gaston and Jones, 1989), so we can only compare weights among years if they are corrected for the date on which they were obtained. When this is done we find that, like egg weight, the mean weight of breeding adults is lower in late years than in early ones (Figure 13.2, weights corrrected to 1 April). Interestingly, there is no similar effect if we compare the weights of birds found in burrows with chicks. It seems that the birds are programmed to build up their weight before incubation, and that the amount that they put on depends on how easy it is to find food. However, even in bad years when laying is delayed, they still manage to finish incubation with about the same level of reserves (Gaston and Jones, 1989). Perhaps the initial weight build-up is an insurance against very bad years; worse than any that occurred during our study.

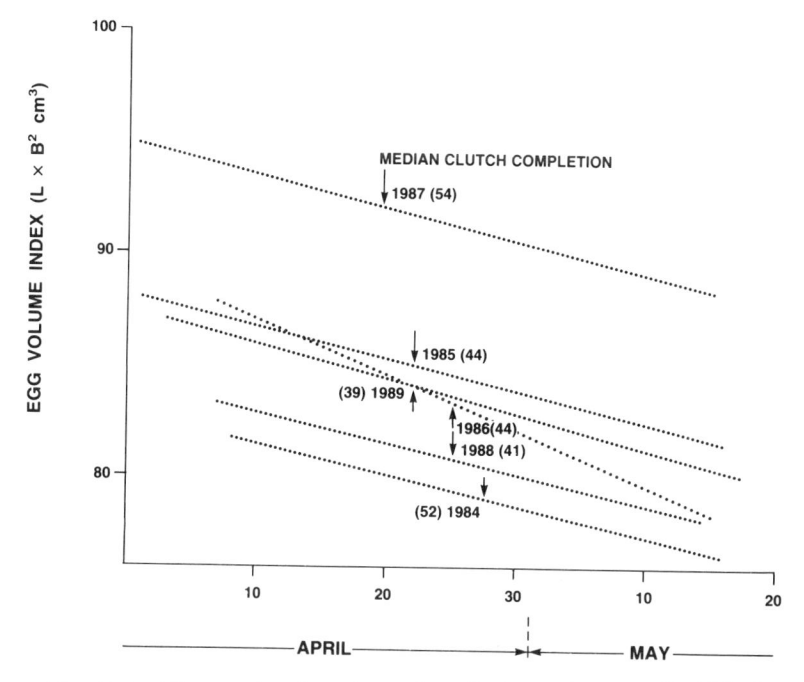

Fig. 13.3 *The regression of egg size on date of laying in six years at Reef Island.*

It would seem axiomatic that the weights of the chicks just after hatching should be closely related to the weight of the eggs, but we found no correlation between mean weights of chicks found in burrows and either mean egg weights, or median dates of laying. However, the mean weights of the much larger sample of chicks captured at departure were correlated with mean fresh egg weight ($r = 0.84$, $n = 5$, $P < 0.05$) and with median date of laying (Figure 13.2). Perhaps these relationships were obscured for the chicks weighed in burrows, by different lengths of time that had elapsed from hatching.

In most years, chicks captured at departure became lighter, on average, as the season progressed (Figure 13.4). The seasonal decline in the weight of chicks was similar, in most years, to the observed seasonal decline in the weight of eggs, both amounting to about 10% of mean weights at the start of the season. However, the decline became steeper towards the end of the departure period in several years. As we saw in Chapter 11, the weight of chicks at departure is strongly affected by the length of time that they have spent in the burrow. The relatively lighter weights of chicks leaving towards the end of the season may indicate that some of these chicks spent an unusually long time in the burrow before departure. The small number of breeders involved may have been less competent, perhaps because less experienced than early breeders.

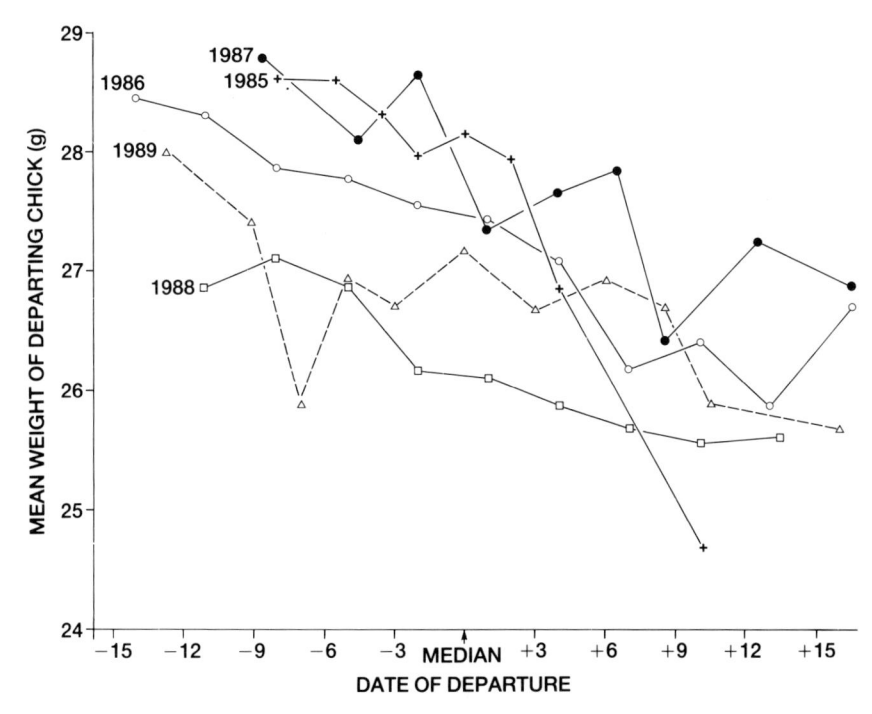

Fig. 13.4 *Weights of chicks at departure, in relation to date.*

POSSIBLE CAUSES OF INTER-YEAR VARIATION

The inter-relationships between date of laying, egg size, and adult and chick weights strongly suggest inter-year differences in the availability of food, at least during the pre-laying period. In the three earliest years, 1985, 1987 and 1989, there were large numbers of other seabird species feeding in Hecate Strait throughout the breeding season (Figure 7.3), and more large whales were present than in other years. We found that the aggregations of Sooty Shearwaters and Black-legged Kittiwakes, and probably also the Humpback Whales, were associated with large swarms of euphausid crustacea close to the surface. It seems likely that the abundance of euphausids was responsible for the feeding concentrations of seabirds and whales, and also for the early laying of the Ancient Murrelets.

Unfortunately, few systematic observations on the oceanography of Hecate Strait are made every year, so it is hard to identify what aspects of the marine environment might have caused the increased abundance of euphausids in certain years, with the attendant effects on seabirds. When I examined the weather records from Sandspit airport, the nearest weather station to Reef Island, I found that there was a correlation between the mean minimum temperature in March (the pre-laying period) and the median date of laying (Figure 13.5). The colder the mean minimum temperature, the earlier the Ancient Murrelets laid. This result seems paradoxical when we consider the effect of temperature over the species' whole range, as described in Chapter 4. However, in the general pattern of circulation, cold water from Queen Charlotte Sound finds its way northward up the west side of Hecate Strait. Low temperatures at Sandspit in spring may relate to an intensification of this cold water inflow, perhaps increasing productivity in Hecate Strait, and causing the various biological effects that we observed. This is pure speculation, but the effect seems to be one worth investigating.

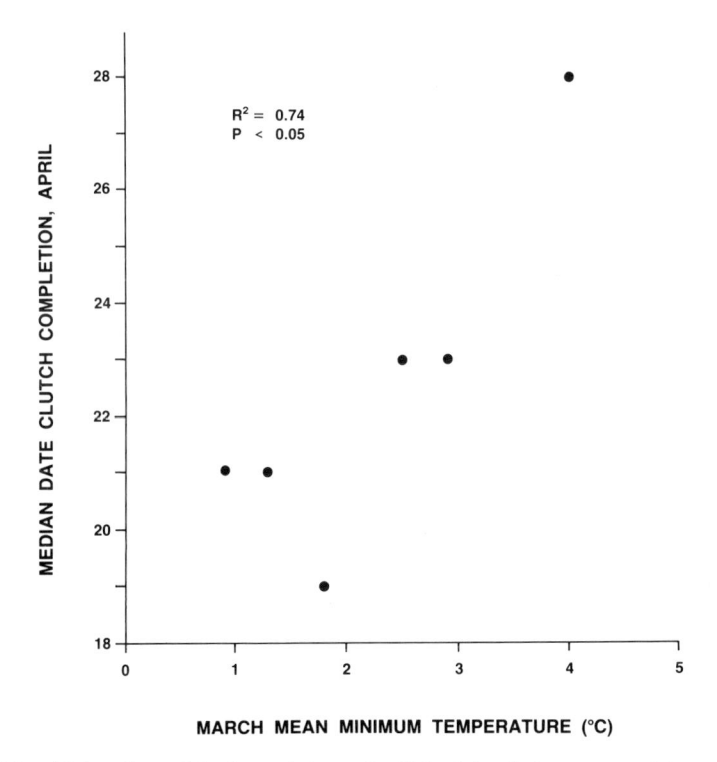

Fig. 13.5 *Median dates of clutch completion at Reef Island, in relation to mean minimum March temperatures.*

INDIVIDUAL VARIATION

In many species, those individuals with little, or no, experience of breeding tend to lay later in the season than more experienced birds (Ryder, 1980). Because we did not know the history of most of our breeders, we could not investigate this phenomenon in the Ancient Murrelet. However, we could compare the timing of breeding for individual birds in successive years, by comparing the dates on which they were captured in the burrow with their chicks. Because timing of laying varied from year to year, we had to express all dates as deviations from the median date of chick departure.

In most comparisons between different pairs of years we found that individual birds tended to be consistently either early, or late. Consequently, there was a correlation between the dates of capture in any two years (Figure 13.6).

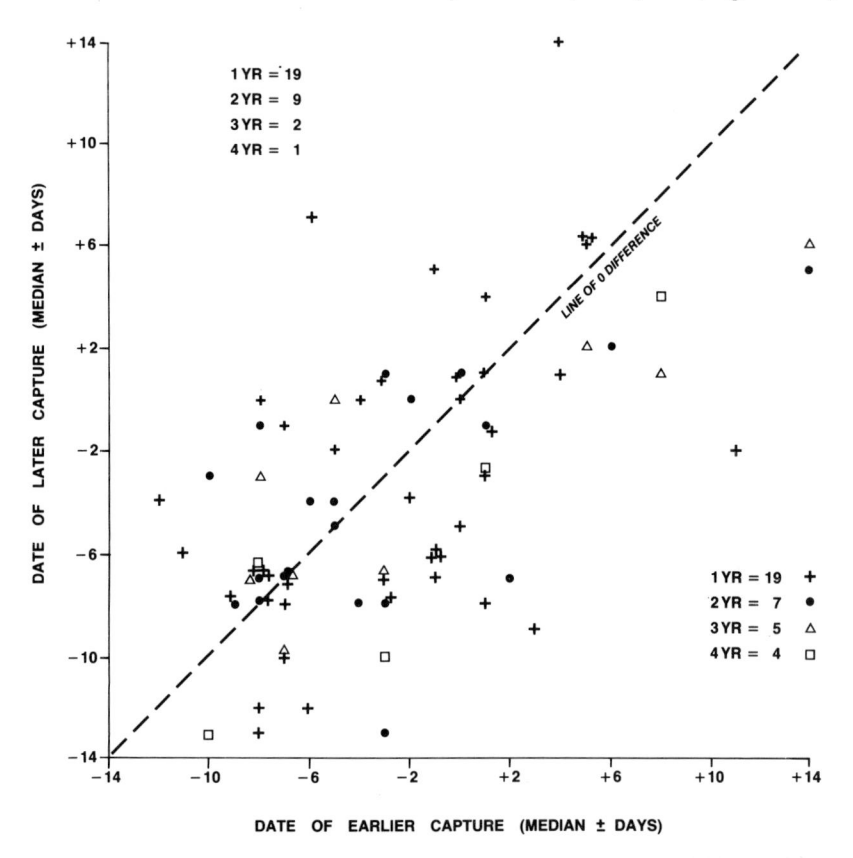

Fig. 13.6 *Dates of capture of adults with chicks, in different years, in relation to the median departure dates for each year.*

When the data from all years were combined, the date of capture in one year explained about 36% of the variation in the date of capture in the other years (Table 13.2). In addition, there was a tendency for dates to become slightly earlier year by year. Birds caught in successive years averaged half a day earlier in the second year than in the first, while those caught 4 or 5 years later averaged nearly three days earlier (Figure 13.7). Presumably, this relates to their increasing experience as breeders. We also examined the weights of birds

Table 13.2 *Correlation coefficients for the date of capture of individual breeders in burrows with chicks in different years (sample sizes in parentheses).*

Year of capture	Year of subsequent capture			
	1986	1987	1988	1989
1985	0.00 (5)	0.57 (11)	0.88 (6)	0.92 (5)
1986		0.70 (8)	0.98 (4)	0.79 (4)
1987			0.86 (8)	0.69 (4)
1988				0.42 (20)

All years combined: $r = 0.60$.

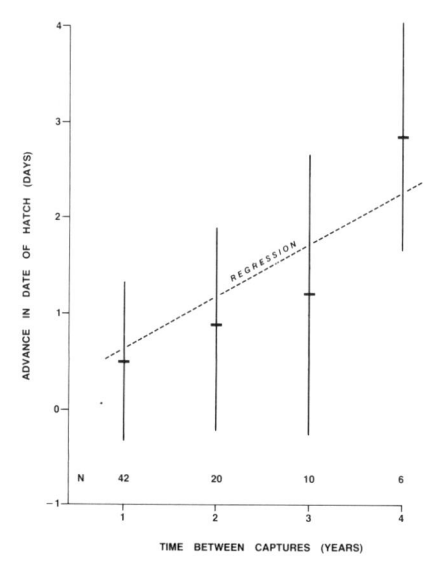

Fig. 13.7 *Changes in the date at which individual birds were caught with chicks (days ± the median for that year) in relation to the number of years that had elapsed between first and second captures.*

captured with chicks, to see if those birds known to have bred for several years were heavier than they had been when first captured, but we found no effect. This agrees with our finding that there was no difference among years in the weights of adults with chicks. It seems that weights just before the departure of the chicks are insensitive to environmental conditions, or the experience of the birds, perhaps because by that stage in the season food is very abundant. However, there was a significant tendency for the weights of individual birds to be similar from year to year. Individual differences accounted for 63% of total variation in the weights of birds captured with chicks ($F_{3,14} = 2.42$, $P < 0.05$).

Although the size of eggs laid is sensitive to variations in environmental conditions, and seems also to be affected by date of laying, these factors explain only a small proportion of the variation among individual clutches. Because we identified only a small proportion of breeding adults in any one year, and because we did not know the sex of some of them, we could obtain only indirect information on the extent to which the size of the eggs was determined by the peculiarities of individual females. To examine this effect, we compared the size of eggs laid in the same burrow in different years, where we knew that at least one member of the pair was the same in both years. The results suggest that females tend to lay similar sized eggs each year, and that variation between different individuals explains much of the difference within years (Table 13.3).

Table 13.3 *Correlations between the volume index ($L \times B^2$) of eggs laid in the same burrow in different years, where at least one member of the pair was the same.*

	Subsequent years			
Year 1	1986	1987	1988	1989
1985	0.37 (10)	0.72 (6)	0.56 (6)	—
1986		0.83** (11)	0.37 (8)	0.38 (4)
1987			0.78* (9)	0.11 (5)
1988				0.27 (15)

* $P < 0.05$, ** $P < 0.01$.
Analysis of variance showed that 47% of variation in egg volumes was accounted for by the presence of at least one common parent in the burrow concerned ($F_{3,13}$, $P < 0.05$).

SUMMARY

Compared to many seabirds that have been studied over several years, the Ancient Murrelets breeding at Reef Island showed only small variations among years in their breeding biology. However, despite the narrow range of median laying dates recorded, the timing of laying seemed to have repercussions for egg size and adult weight. These variations synchronized, to some extent, with variations in the abundance of certain non-breeding populations of seabirds in Hecate Strait, which may have been caused by variations in the availability of euphausid crustacea. Hence, some features of Ancient Murrelet breeding biology which can be easily measured, such as timing and egg-size, may be good indicators of important events in the marine environment at large.

In any one year, most of the variation within the colony in timing of laying and egg size was determined by characteristics of the individual breeders, although there was a slight tendency for birds to lay earlier in the season, relative to the median for the population, as they got older. Birds laying right at the end of the season may be somewhat less competent than those laying earlier, as evidenced by the smaller eggs and lower weights of their chicks at departure. Some of these birds may be first-time breeders. However, there is little evidence that experience has a strong effect on reproductive performance.

CHAPTER 14

Population dynamics

How we distinguished breeders from non-breeders; the age of nonbreeders and when they first start to breed; movements within the colony; adult survival; reproductive success; predation by mice; the survival of chicks during departure; the age structure of the population; movements between colonies.

Finding out about the population dynamics of Ancient Murrelets was one of the main objectives of our work on Reef Island. Little was known on the subject before we started, mainly because hardly any Ancient Murrelets had been banded previously. A few educated guesses had been made, mainly by analogy with other, better known, species of small auks (e.g. Sealy, 1976; Nelson, 1977). Otherwise, the only definite information was an estimate of reproductive success by Vermeer and Lemon (1986).

Finding out about the population dynamics of Ancient Murrelets was important for two reasons. Firstly, if we want to understand the effects of unnatural mortality from oil spills, or introduced predators, we need to know what the normal rates of reproduction and mortality are. Secondly, theoretical treatments of life history tactics suggest that different demographic parameters should be interrelated (Williams, 1966; Goodman, 1974; Stearns, 1976). A trade-off between fecundity and adult survival has been demonstrated for North American game birds by Zammuto (1986), and for all birds by Gaillard *et al.* (1989). If the Ancient Murrelet differs from other alcids in its demographic characteristics, then these demographic differences may have been involved in the evolution of the precocity in the chicks. Hence, knowledge of the Ancient Murrelet's demography, and comparison with other, non-precocial, auks, might provide us with clues about its peculiar life-history strategy.

Once, when I presented our results on the population dynamics of Ancient Murrelets at a scientific meeting I subtitled my talk, "what we learned in six years of tramping about in the forest in the middle of the night". The results do not seem all that spectacular, considering the labour that went into them. Nevertheless, understanding demography is the key to making reasonable decisions about population management. If we are not prepared to do the work necessary to obtain at least crude estimates of reproductive rate and survival, then probably we should not try to pretend that we can predict the effects of environmental changes on populations. Demographic studies of single species may not make very exciting science, but they are basic building blocks without which practical ecologists cannot get very far.

AT WHAT AGE DO BIRDS BEGIN TO VISIT THE COLONY, AND HOW OLD ARE THEY WHEN THEY BEGIN TO BREED?

Answering these questions was one of the main reasons for our large-scale banding of chicks. Initially, I had assumed that most chicks would return to breed on their natal island. Hence, I anticipated that, after several years of banding, about 20% of the non-breeders that we trapped would be banded. Unfortunately, things did not work out that way. In 1989, after four seasons of intensive banding, only 13/412 (3%) non-breeders captured on Reef Island had been banded as chicks. This suggested strongly that our chicks were mixing with a much larger population, perhaps originating from most of the colonies on the east coast of Moresby Island. While this was an interesting finding in its own right, it meant that our samples of known age birds were much smaller than we had hoped.

By the end of 1989 we had caught two 1-year-old (1Y), eighteen 2-year-old (2Y), six 3-year-old (3Y) and one 4-year-old (4Y) birds on Reef Island (Table 14.1). In addition, we caught three 2Ys at Limestone Island in 1989, the only

Table 14.1 *Numbers of chicks banded and numbers retrapped in subsequent years.*

Year banded	Total banded	Year recaptured			
		1986	1987	1988	1989
1985	328	1	4	3	1
1986	720		0	5	3
1987	1070			1	12
1988	1256				0

birds banded at Reef Island to be retrapped there, out of 250 captured. They provided further evidence that non-breeding birds visiting a given colony included a substantial number originating from elsewhere.

Two of our birds of known age appeared to be breeding when recaptured, a 2Y caught on 23 May 1989 with a complete brood-patch, and a 3Y caught on 22 May 1988 without a brood-patch and again on 6 June 1988 with a fully-developed brood patch. The latter bird had apparently initiated breeding very late in the season, as might be expected for a first-time breeder (Ryder, 1980). It suggested that a few birds, identified by our criteria as non-breeders, actually may have bred. Otherwise, among birds of known age, only the single 4Y that we captured showed any signs of a brood-patch (an incomplete one 10 mm across). A 3Y, trapped on 7 April 1988, qualified as a breeder based on the date of capture. Apart from this bird, the earliest date of capture of a bird banded as a chick was 12 May.

We recaptured 0.1% of the 1Ys that we banded ($n = 3374$), 1.0% of 2Ys ($n = 2118$), 0.6% of 3Ys ($n = 1048$) and 0.3% of 4Ys ($n = 328$). Recapture rates for 2Ys varied from 0.7% for the 1986 cohort in 1988 to 1.2% for the 1985 cohort in 1987. The average recapture rate over the three years available was 1%. These figures suggest that few 1Ys visit a colony, but most 2Ys do so. The lower recapture rates of 3Y and 4Y birds may relate to our general observation that non-breeders arriving in May were easier to catch than birds arriving earlier in the year. The older age classes may have been more experienced on land and therefore less likely to be caught than 2Ys.

Some clues about the age of most non-breeders can be derived from weights and measurements. When we compare the weights and wing lengths of breeders and non-breeders, we find that they differ considerably, with non-breeders being lighter, and having shorter wings (Table 14.2). The weights and wing-lengths of non-breeders were much closer to those of 2Ys than 3Ys, and this suggests that 2Ys predominated among the non-breeders of unknown age.

Table 14.2 *Weights and wing lengths of birds captured at Reef Island.*

	Weight			Wing-length		
Status/age	Mean	S.D.	n	Mean	S.D	n
Breeders (2nd half of May)				140.2	3.2	533
Non-breeders	184.6	11.6	924	138.1	3.2	859
2Ys	180.1	7.4	20	137.7	3.1	21
3Ys	197.5	13.6	4	139.3	2.8	6

In most auks for which information is available, non-breeders return to the colony earlier in the season as they get older (Hudson, 1985). The distribution of recaptures of 2Y and 3Y birds over the course of the season is very similar to that for all non-breeders (Figure 14.1). Consequently, there is no evidence to

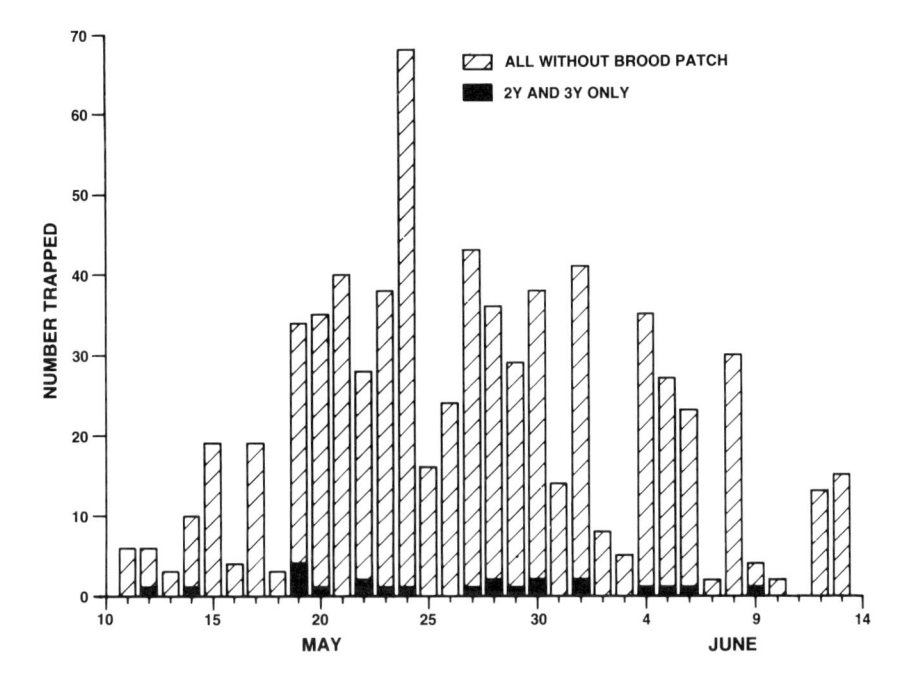

Fig. 14.1 *The date of trapping of second and third year birds, and of all birds without brood patches, at Reef Island.*

suggest that substantial numbers of non-breeders older than three years old attend the colony.

We can get some idea about how many seasons birds attend the colony as non-breeders before they begin to breed by looking at what proportion of birds trapped as non-breeders were breeding when retrapped in subsequent years. Out of 31 birds trapped as nonbreeders and retrapped the next season, 21 (68%) were breeders when retrapped. Out of 11 retrapped two years later, all but one (91%) were breeding. The corresponding figures for birds trapped on the surface as breeders were 90/95 (95%) and 60/61 (98%). The high proportion of birds retrapped as breeders one year after being trapped as non-breeders suggests that the majority of Ancient Murrelets visit the colony for only one or two years before they begin to breed. Considering that some failed breeders may have been classified as non-breeders, the proportion of birds trapped as breeders in consecutive years makes it seem likely that Ancient Murrelets, once they have begun breeding, only visit the colony in subsequent years if they are going to breed. However, it does not rule out the possibility that some birds might miss a year of breeding, if they visit the colony little or not at all while doing so.

Taking all the evidence together, it appears that birds in their first summer visited the breeding colonies infrequently, and that the majority of non-breeding birds trapped were probably in their second summer. Most birds visited the colony for only one or two seasons before they began to breed. However, some birds bred in their second summer, while some third- and fourth-summer birds were still non-breeders when trapped, so there was some variation in the age at which breeding began. Probably most birds are breeding by the time they reach their fourth summer.

MOVEMENTS WITHIN THE COLONY

We usually recorded what part of the colony each bird was trapped in, and we used this information to assess how much birds moved about, both before and after starting to breed. Of birds banded when fully grown, 90% ($n = 286$, excluding birds removed from burrows) of retraps were in the same area where they had been banded. The proportion of breeders retrapped away from their area of banding (8%, $n = 235$) was much lower than the corresponding proportion of non-breeders (22%, $n = 51$). If we consider only the number retrapped in areas not contiguous with the area in which they were banded, the difference is greater (breeders 1%, non-breeders 14%). At two catching areas separated from the others by more than 200 m (F and J), no birds banded elsewhere were retrapped among 106 birds captured there. Only one bird banded on Reef Island when fully grown was retrapped on Limestone Island—a non-breeder banded in 1989 and retrapped, again as a non-breeder, in 1990. We can conclude from this that most Ancient Murrelets remain very close to their breeding site, once they have selected it. This observation makes it unlikely that estimates of survival rates are much influenced by emigration to other parts of the colony.

ANNUAL SURVIVAL

Adult survival was estimated from retrapping data using the regression method (Furness, 1978; Kampp, 1982), and a computer program (SURGE) developed by Clobert and Lebreton (1985). The program allows survival to be estimated separately for different years after capture, and provides estimates of how the probability of recapture varies among years (Lebreton and Clobert, 1986). Banding and retrapping data on which the survival analyses were based is given in Table 14.3. The regression of the proportion of breeders ($\log_{10}$) retrapped on the number of years elapsing since banding (Figure 14.2) gives an estimate of 75% annual survival, with 95% confidence interval 69–82% (Table 14.4). If only birds trapped in burrows are considered, the estimated survival is 67%, and if all retraps are considered, 84%.

The SURGE estimate based on the figures in Table 14.3 gives an annual survival of 69%, if inter-year variation is ignored. If survival is estimated separately for the first year after capture and for subsequent years the estimates are 47% and 77% (Table 14.5). The difference between the first and subsequent years after trapping suggests that the chance of retrapping a bird in the first year after banding was lower than in subsequent years, perhaps because the disturbance caused by capture affected the probability that an individual would visit the colony the following year. The SURGE estimate for survival, excluding the first year after capture, is very close to that obtained by the regression method.

The SURGE program also estimates the probability of recapture by year. These probabilities varied from 11% in 1986 to 49% in 1985, with probabilities for the other three years falling between 25–33%. These inter-year differences in recapture probabilities reflect changes in catching operations. In 1984 and 1985 practically all trapping was confined to one small area, giving the high recapture rate observed. In 1986 the catching area was expanded and trapping in the previous banding area was less intensive. In 1987 the area was again expanded, but with increased effort overall, after which our catching area and effort remained fairly similar.

Table 14.3 *Years of banding of birds retrapped in subsequent years.*

Status	Year banded	Total banded	1985	1986	1987	1988	1989
All breeders	1984	125	27	4	0	5	6
	1985	137		7	18	16	10
	1986	124			18	16	17
	1987	132				16	17
	1988	274					36
Breeders in	1984	17	1	1	0	0	1
burrows only	1985	26		5	11	7	6
	1986	15			4	1	0
	1987	20				4	5
	1988	25					11
Non-breeders	1984	32	3	1	1	1	0
	1985	78		2	1	3	4
	1986	70			3	3	2
	1987	289				13	7
	1988	313					12
All trapped	1984	188	35	5	2	8	8
as adults	1985	246		9	19	22	14
	1986	172			21	19	18
	1987	443				31	26
	1988	588					52

204 Studies at Reef Island

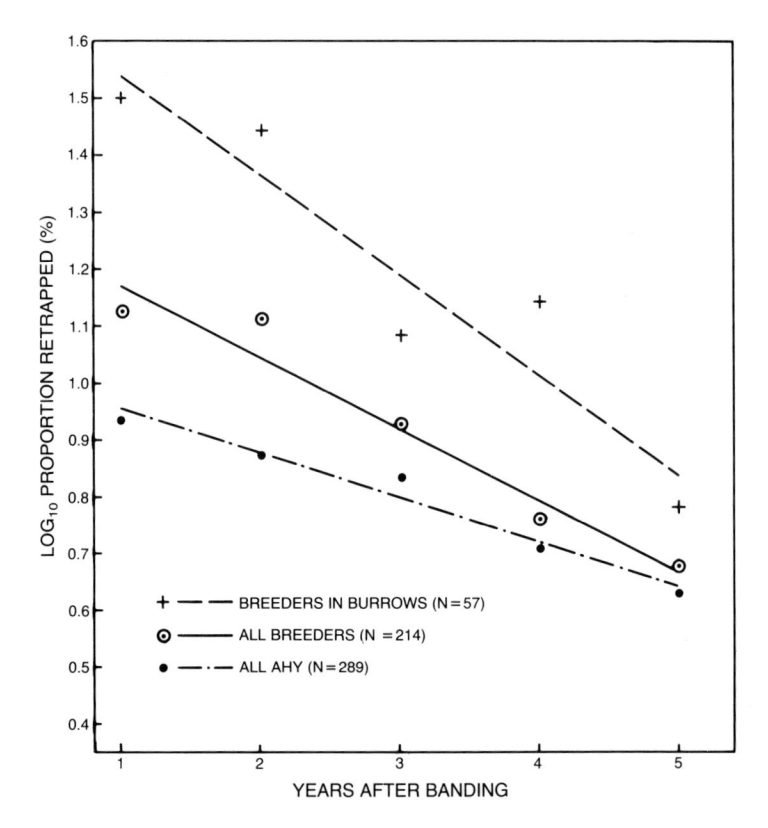

Fig. 14.2 *Proportions (log) of Ancient Murrelets retrapped at Reef Island, in relation to the length of time elapsed since their first capture.*

Table 14.4 *Regression estimates of adult annual survival.*

Sample	Regression slope (±95% CONF. INT.)	Annual survival (±95% CONF. INT.)	R^2
ALL RECAPTURE YEARS			
All breeders	−0.123 (0.087–0.163)	0.75 (0.69–0.82)	0.95
Breeders in burrows	−0.175 (0.090–0.259)	0.67 (0.55–0.82)	0.88
All captures	−0.078 (0.057–0.100)	0.84 (0.77–0.88)	0.96
OMITTING FIRST YEAR AFTER CAPTURE			
All breeders	−0.147 (0.108–0.186)	0.69 (0.65–0.78)	0.97
Breeders in burrows	−0.194 (0.054–0.334)	0.64 (0.46–0.88)	0.84
All captures	−0.086 (0.054–0.118)	0.82 (0.76–0.88)	0.95

Table 14.5 *SURGE estimates of adult survival and recapture probabilities (version SURGE 4.0).*

Parameter	Est. value	Min.	Max.	S.E.
SURVIVAL				
Breeders				
First year	0.474	0.365	0.585	0.057
Other years	0.767	0.666	0.845	0.046
All years combined	0.691	0.612	0.759	0.038
All birds captured				
First year	0.355	0.284	0.432	0.038
Other years	0.819	0.720	0.889	0.043
RECAPTURE PROBABILITIES, BREEDERS				
1985	0.494	0.310	0.680	0.099
1986	0.105	0.056	0.188	0.033
1987	0.247	0.171	0.342	0.044
1988	0.303	0.217	0.404	0.048
1989	0.327	0.238	0.429	0.049

REPRODUCTIVE SUCCESS

Because of the disturbance caused by our inspections, we could not use any of the information collected in the early years to estimate reproductive success. However, in 1988 and 1989, when we used the temperature probes to study incubation without disturbing the birds, we believe that our activities had little effect on the birds. Information from other years could be used to estimate clutch size, and the survival of chicks at departure.

Nearly all clutches for which incubation was begun were of two eggs (286/298, 96%; Table 14.6), and most broods consisted of two chicks at departure (154/168, 92%). Out of 336 eggs incubated for at least 30 days, 322 (96%) hatched. Three desertions occurred after hatching, but these were probably due to our interference (Gaston and Powell, 1989). Under natural conditions practically all chicks that hatch apparently survive to depart from the burrow in the normal way.

In 1988 and 1989, 76% and 88% of pairs respectively succeeded in departing with at least one chick. Most failures involved desertions before incubation had begun (10–28% of burrows in different years). Such desertions accounted for 8/10 failures in 1988 and 5/6 in 1989. We assumed that these were not connected with our activities, because the birds never encountered us, but the possibility that some signs of our visits had put the birds off cannot be entirely discounted. Overall reproductive success was 1.46 and 1.60 chicks per pair (Table 14.6).

Information on reproductive success is available from two other studies. At Frederick Island in 1980 and 1981 Vermeer and Lemon (1986) recorded

Table 14.6 *Reproductive success of Ancient Murrelets at Reef Island.*

Parameter	1984	1985	1986	1987	1988	1989	Totals
Clutch							
1	4	0	1	0	1	2	8
2	47	61	45	47	40	46	286
3/4	0	2	0	0	0	2	4
Mean	1.92	2.05	1.98	2.00	1.98	2.00	1.99
Deserted without inc.	6	6	—	13	8	5	38
% deserted	12	10	—	28	20	10	15
Brood at departure							
1	0	0	2	2	2	8	14
2	9	31	22	27	29	36	154
Mean	2.00	2.00	1.92	1.93	1.94	1.82	1.92
Eggs							
Incubated >30 days	20	62	48	58	62	86	336
Hatched	18	62	46	56	60	80	322
% Hatched	90	100	96	97	97	93	96
Chicks departing/pair	—	—	—	—	1.46	1.60	1.54

breeding success on the basis of inspections made only late in the incubation period. Their study pairs reared 1.48 and 1.69 chicks/pair to departure in the two seasons. These figures may exaggerate reproductive success slightly, because they presumably omit any burrows where eggs disappeared early in incubation (only a few at Reef Island). In 1982 Rodway and Lemon (in Rodway *et al.*, 1988) made a small number of visits to 25 burrows at Dodge Point, Lyell Island, and found an overall reproductive success of 1.44 departing chicks/pair. In both studies, as at Reef Island in 1988 and 1989, most failures occurred because incubation was never initiated. These results confirm that, under undisturbed conditions, Ancient Murrelets in the Queen Charlotte Islands rear to departure about 1.5 chicks per breeding pair per year.

PREDATION BY DEER MICE

In all years, when we inspected the burrows, we found a few eggs and chicks that clearly had been partially eaten by mice. In 1984 and 1985, when there was much desertion as a result of our interference, we found that if we left deserted eggs in the burrows most were eaten by mice eventually, although many unincubated or partially incubated eggs survived for several weeks. However, if the eggs were pipping when deserted, as occurred occasionally, they were invariably attacked within one or two days. We did not observe any instances where a first egg was eaten within eight days of laying (the usual interval between layings), except for two cases where the eggs were already cracked. These observations suggested, either that the mice did not recognize the eggs as a source of food unless they were cracked open or that they were unable to break open intact eggs.

We placed several mice in plastic buckets with intact Ancient Murrelet eggs and no other source of food. Under those conditions they did not break into the eggs. Further evidence that they cannot open intact eggs came from our experiments with tea-stained chicken eggs (Chapter 12). In some cases, after the eggs had been in the nest for a few days, we could clearly see the marks of mouse incisors, where they had scraped away the staining, without breaking the shell. We concluded that mice had considerable difficulty in breaking open murrelet eggs. Hence, it seems likely that most predation by mice at Reef Island simply removes eggs that are already deserted, or cracked. The chicks are clearly vulnerable to mice, but they are normally brooded by one of their parents throughout their stay in the burrow. Judging from the ferocity with which some birds attacked our hands, the breeders should be fully capable of repelling mice.

During their censuses of Ancient Murrelet colonies throughout the South Moresby area, Rodway *et al.* (1988) recorded all remains of eggshells that they encountered and estimated the amount of egg predation on that basis. On many colonies the number of eggs apparently taken amounted to a sizeable fraction (up to 30%) of those estimated to have been laid. However, they did not distinguish between those that had been eaten by predators and the shells of eggs which had hatched. On Reef Island we found that eggshell remains were frequently dragged out of the burrow during family departures, providing a good clue that the brood had gone. Consequently, the impact of egg predation on Ancient Murrelets is probably much lower than it would appear from the figures of Rodway *et al.*

Murray *et al.* (1983) found that predation by deer mice on Xantus' Murrelets' eggs was very heavy, with much of it occurring before incubation was initiated (28% of 470 eggs were taken at that stage). The eggs of Xantus' Murrelet, a smaller bird, are smaller than those of the Ancient Murrelet. In addition, the nest sites are often in rocky ground, where the eggs may be easily cracked, if the mice roll them about. At Reef Island we noted that eggs in rocky burrows were more likely to be cracked than those in earth burrows. The differences in size and nesting sites may account for the much higher predation on the eggs of Xantus' Murrelets.

BROOD SIZE AFTER LEAVING THE COLONY

There are several obstacles that may reduce the numbers of chicks immediately after they leave the burrow. On Reef Island chicks are taken by Saw-whet Owls and deer mice during their trek from the burrow to the beach

(Chapter 8). A few may also be held up by becoming trapped in difficult terrain. On Limestone Island Moira Lemon and Andrea Lawrence found an entire family trapped in a deep hole, from which they had to be released. Chicks are also eaten by large fishes close to the beach. To see what effect these sources of mortality had on the population, we attempted to find families at sea on the morning after their departure.

By the time that it becomes light most family parties are already several kilometres from the colony. They have also scattered over a wide area, and this makes it very hard to find them (cf. Sealy, 1976). We attempted to locate family parties at dawn in all years except 1984, finding ten in 1985, four in 1986 and 1987, and two in 1988 and 1989. All but three parties included two chicks, one party was of three, and the other two were of one. The group of three chicks was probably the result of at least one chick attaching itself to the wrong parents. The proportion of parties including two chicks (86%) is similar to that at departure from the burrow (92%), suggesting that losses between the chicks leaving the burrow and making a rendezvous with their parents are relatively low.

Two other studies mention the size of broods of young Ancient Murrelets soon after departure from the colony, although neither attempted to count them on the first day. Guiget (1953) saw eight family groups in Queen Charlotte Sound on 13 June, each including two young. At that date the chicks could have been several weeks old. Sealy and Campbell (1979) reported ten families off the Queen Charlotte Islands, eight of two and two of one chick. Assuming that broods do not amalgamate, these observations suggest that chick survival immediately after departure from the colony is high.

AGE STRUCTURE OF THE POPULATION

We can use the estimates that we have obtained of reproductive success and survival to calculate the age structure of the population, and the proportion of birds attending the colony in a given year. If we assume that half of the population first breeds at 3 years, and the rest at 4 years, that all continue to breed annually thereafter, and that the survival of non-breeders from the age at which they return to the colony is similar to that of breeders, we can estimate the survival of chicks from departure to return at 2 years old (S_2). Using the estimates presented here for the annual survival of breeders (S_b, 75%), and the annual production of young at departure from the colony (r, 0.77 chicks per female breeder), gives

$$S_2 = \frac{2(1 - S_b)}{r\, S_b\, (1 + S_b)} = 0.49$$

To estimate the age structure of the population during the breeding season, and hence what proportion of the population is actually breeding, I have assumed that half of the mortality occurring between departure and 2 years (using the value of 49%) happens before the first spring. In that case, at the start of breeding, there will be 0.54 1Ys, 0.38 2Ys, and 0.14 non-breeding 3Ys for every breeder of the previous year, of which 0.75 remain. Hence 30% of the population is made up of 1Ys which do not visit the colony, 29% of non-breeding 2Ys and 3Ys, and 41% of breeders 3 or more years old. If the mortality of pre-2Y birds is concentrated mainly in the period immediately after leaving the colony, which seems likely, the proportion of 1Ys may be lower, but the ratio of non-breeders to breeders among birds visiting the colony remains unaffected.

MOVEMENTS AMONG COLONIES

The capture on East Limestone Island of several birds banded at Reef Island as chicks shows that some non-breeders prospect a variety of colonies before deciding where to breed. The proportion of banded birds among non-breeders trapped at Reef Island supports this. The census of burrows, and the number of chicks trapped, both suggested that about 534 pairs bred (P_b) within the catchment area of the chick trapping funnels. In 1989 12 birds banded as chicks in 1986 and 1987 were recaptured on Reef Island (C), out of 595 non-breeders (NB), presumably mainly 2Ys and 3Ys. We can estimate the total number of pairs involved in producing the non-breeders (N_p) from which the Reef Island sample was trapped (the source population) as follows

$$N_p = \frac{P_b \times NB}{C} = 26\,477 \text{ pairs}$$

The nearest colonies to Reef Island are those on the Limestone islands (*c.* 1000 pairs, 6.5 km away), and at Dodge Point, Lyell Island (*c.* 11 000 pairs, Rodway *et al.*, 1988), about 15 km away. Otherwise, the nearest colonies are more than 30 km away, at Ramsay, House and Agglomerate islands in Juan

Perez Sound. The estimated size of the source population supplying potential recruits to Reef Island suggests that population exchange may extend at least as far as Juan Perez Sound. This has implications for their conservation, because it may enable Ancient Murrelet populations to shift between colonies through the differential establishment of prospectors. Conversely, some colonies could persist in the face of very heavy predation, and act as population sinks, something which appears to be happening at the Limestone islands, where predation remains amounted to more than 50% of the estimated breeding population in 1989 (pers. obs.).

SUMMARY

Most birds begin to visit the colony in their second summer, and breeding probably begins at three or four years old. Breeders are almost always retrapped again in the area where they were caught, but those trapped as non-breeders may be caught elsewhere, showing that they prospect different parts of the colony. About 75% of breeders survive from one season to the next, and each pair produces 1.5 chicks per year, on average. Losses during departure from the colony appear to be low. Using the estimates for adult survival and numbers of chicks produced, we can calculate that about 50% of the chicks leaving the colony survive to the age of two years. This means that the population during the breeding season must comprise about 30% first-years, which do not visit the colony, 29% non-breeding 2- and 3-year-olds, and 41% breeders of 3 or more years old. The very low proportion of banded chicks recovered at their colony of origin suggests that young birds prospect many colonies and that many settle away from their natal island.

CHAPTER 15

Why Ancient Murrelet chicks are precocial

> *Difficulties involved in interpreting complex traits, such as life-history adaptations; what the ancestors of the Ancient Murrelet probably did; previous theories, and why I do not find them satisfactory; a new theory to account for precocity.*

For the most part, the studies that we carried out at Reef Island were not designed to determine the causes of the Ancient Murrelet chicks' unusual departure strategy. However, the information that we obtained did allow us to evaluate possible hypotheses in the light of better data than were available previously. In this chapter I shall review some of the ways in which Ancient Murrelets differ from other members of their family, and discuss how these differences might have contributed to the evolution of precocity. I shall also discuss some previous ideas on why the chicks are precocial, and put forward my own suggestion.

The problem with discussing the evolution of a major life-history adaptation, is that it usually involves the adjustment of many different features of the animals' biology. Many of these traits could not have evolved without some, or all of the others. For instance, Ancient Murrelets breed on islands where there are large number of day-hunting predators: eagles, falcons, ravens and crows. To avoid these predators, the murrelets nest far underground and visit the colony only when it is completely dark. To breed under these conditions, they have developed a very complex array of calls and a display area at sea (the gathering ground) to cope with the difficulties of locating and communicating with mates. Those posture and plumage displays that they maintain must be designed to have effect at sea rather than on land. At the same time, during the breeding season, their opportunities for feeding at night are reduced, presumably causing selection for more efficient day-time foraging, and perhaps affecting the consequent evolution of their bills and digestive systems.

Selection for a complex of traits will almost certainly involve a mixture of many selective forces, and the relative importance of different forces may differ depending on the particular trait involved. With many contributory causes, and many interrelated traits, the process of identifying a chain of cause and effect becomes very difficult, especially as we have no way to test our ideas experimentally. Consequently, we are hardly likely to be able to reach a firm conclusion about why precocity evolved in the Ancient Murrelet. The best that we can do is develop an explanation that accords reasonably well with what we know of the species' recent ecology, and hope that things have not changed too much during its evolution.

CHICK REARING IN SEABIRDS

All alcids, except for *Synthliboramphus* and the inshore-feeding *Cepphus* species, have a clutch size of one (Lack, 1968; Sealy, 1973b). The same applies to nearly all other flying, offshore feeding seabirds, because the time taken in travelling to distant feeding areas precludes delivering food at a rate sufficient to provision more than one, normally semi-precocial, chick during the period of maximum demand (Ashmole, 1963; Lack, 1968; Goodman, 1974; Asbirk, 1979; Ricklefs, 1983; Pennycuik *et al.*, 1984).

Among birds that lay clutches of more than one egg, there is generally some variation in clutch size. It is easy to envisage how a reduction in clutch size could evolve, through selection of those individuals which laid the smallest clutches. Such selection would lead eventually to a uniform one-egg clutch. Once this state has been reached, however, the process is difficult to reverse, because there is no variation on which selection can operate. Hence, the evolution of an exclusively single-egg clutch can be regarded as an evolutionary one-way street. Because of this, it seems simplest to assume that the *Synthliboramphus* murrelets are descended from ancestors which, like themselves, laid more than one egg.

CHICK DEPARTURE STRATEGIES IN THE AUKS AND THEIR RELATIVES

Most auk chicks are semi-precocial, which means that they are active, and capable of a wide range of movements, but remain in the nest during the growth period (Nice, 1962). The parents feed them at the nest site for from three to seven weeks and the chicks leave fully feathered and weighing at least 50% of the adult weight. Apart from *Synthliboramphus* species, there are two other genera which do not fit this generalization; the murres *Uria* spp., and the Razorbill *Alca* (and probably also the extinct Great Auk; Bengtson, 1984). In the latter group the chicks are fed at the nest site for about 20 days, but only reach about 25% of adult weight. They develop juvenal plumage before departure, except for the flight feathers (primaries and secondaries), which only grow after they have left the colony. Sealy (1973b) called this the "intermediate" departure strategy, to distinguish it from the precocial *Synthliboramphus* and the semi-precocial genera that comprise the rest of the family. The intermediate and precocial strategies are alike in requiring continuing parental care after the chicks have left the colony, although in the intermediates, where only one young is reared annually, only the male performs this task (Harris and Birkhead, 1985). In the semi-precocial species, chicks leave on their own and are independent thereafter.

The closest relatives of the auks are probably the gulls (Sibley and Ahlquist, 1990), all of which have semi-precocial chicks. Added to the fact that most extant auks are semi-precocial, it is reasonable to assume that the ancestral alcid was likewise semiprecocial (as most other seabirds are), and that the intermediate and precocial strategies have evolved from the semi-precocial. This assumption is crucial, because it means that to explain the diversity of departure strategies among alcid chicks, we must look for selection pressures that caused semi-precocial species to become precocial, or intermediate, rather than vice versa. Hence, my ideas about the evolution of precocity in *Synthliboramphus* are based on the assumption that their ancestors, at one time, reared at least two semi-precocial chicks, provisioning them at the nest site.

PREVIOUS EXPLANATIONS FOR CHICK DEPARTURE STRATEGIES IN AUKS

The intermediate species are all fairly large in body size. The Great Auk, which weighed about 3 kg, was historically the largest of the family; its demise leaves that distinction to the murres, at about 1 kg (Harris and Birkhead, 1985). The Razorbill, though smaller than the murres, is nevertheless exceeded in size, among semi-precocial species, only by the Tufted Puffin. Among auks in general, the ability to carry food back to the chick(s) is not correlated with body size (Figure 15.1). Consequently, the larger auks carry smaller meals, relative to their weight, than the smaller species do. There is a consensus among most writers that the poor load-carrying ability of the larger auks, which have very high weight to wing-area ratios, has been a major factor causing the chicks to depart at an intermediate stage of development. It is presumed that the parents are incapable of supplying food fast enough to allow the chicks to grow to more than a quarter adult weight on the colony (Birkhead and Harris, 1985; Gaston, 1985). This idea supposes that the intermediate strategy is an extension of the process which brought about the adoption, by most offshore-feeding seabirds, of a one-egg clutch.

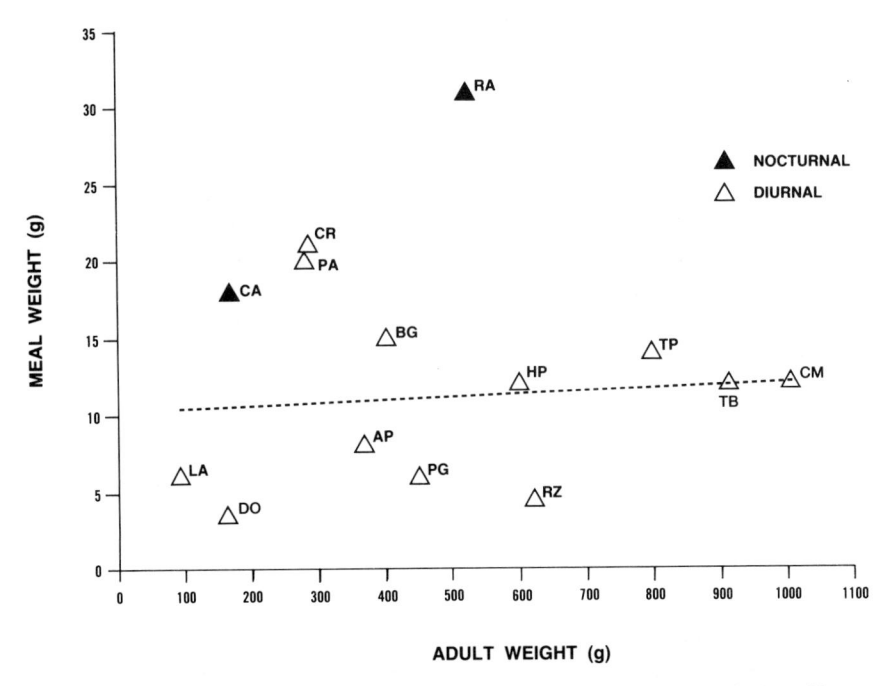

Fig. 15.1 *Mean weights of meals delivered to nestling auks, in relation to adult weights. Abbreviations and sources: CA Cassin's Auklet (Vermeer, 1984), RA Rhinoceros Auklet (Vermeer and Cullen, 1979), CR Crested Auklet (Bedard, 1969c), LA Least Auklet (Roby and Brink, 1986), DO Dovekie, AP Atlantic Puffin (Harris and Birkhead, 1985), HP Horned Puffin, TP Tufted Puffin (Wehle, 1983), BG Black Guillemot (Cairns, 1987), PG Pigeon Guillemot (Koelink, 1972), RZ Razorbill, CM Common Murre, TM Thick-billed Murre (Harris and Birkhead, 1985).*

If we accept that the intermediate strategy evolved because of constraints on the rate at which parents could provision their chicks, there is a temptation to regard the precocial strategy as an extreme example of the same effect. The very term "intermediate" tends to suggest it, although this may not have been Sealy's intention when he coined it. Sealy's (1973b) own hypothesis for the evolution of precocity in *Synthliboramphus* was that the parents were constrained in their ability to provision their chicks by the distribution of their prey. He proposed that the food supply of the Ancient Murrelet was very patchy, and that the patches were hard to locate. He based this theory on his own experience in trying to locate Ancient Murrelets off Langara Island, where he found feeding aggregations to be very unpredictable. Consequently, he supposed that the chicks could be most easily provisioned by leading them to the food source. Sealy (1972) wrote: "The ecological advantage alone of taking the young to the food source adequately accounts for this [precocial] strategy . . .". It would be equally true of any offshore feeding alcid, or for that matter any seabird, that taking the chicks to the feeding area would greatly reduce provisioning journeys. However, that does not explain why it should be *Synthliboramphus* and no other genus which has adopted the strategy.

Two other general theories have been developed about the variety of departure strategies found among auks. Cody (1973) proposed that the amount of time spent on the colony by the chicks was related to their vulnerability to predation. According to his theory, the chicks of murres, and many Razorbills, which are reared on exposed sites, are more vulnerable to predation by gulls than other auk chicks which are sheltered in burrows, rock crevices or other inaccessible sites. Hence, predation by gulls was considered the main selection pressure in the evolution of the intermediate strategy. Clearly this hypothesis does not explain precocity in *Synthliboramphus* because, as Sealy (1973b) pointed out, the murrelets use well-protected sites.

A more recent model was proposed by Ydenberg (1989), combining the idea of predation with that of constraints on chick provisioning. He noted that chicks grew faster on the sea than they did on the colony, and he assumed that the probability of mortality while the chick remained at the breeding site was lower than its chance of dying while on the sea. Hence, he suggested that the chicks leave for the sea when the advantage of the protection afforded by the breeding site (in terms of fitness) is surpassed by the advantage of a higher growth rate on the sea.

In fact, murre and Razorbill chicks leave the colony at the point when they are no longer increasing in weight (Gaston, 1985). Clearly there would be no future in the chicks remaining indefinitely in a situation where they could not grow any further. This point was also made by Birkhead (1977). Ydenberg's theory provides a model for the intermediate departure strategy but does not explain the precocial strategy unless we suppose that, right from hatching, the chances of the precocial chick surviving are higher at sea than on land. There is no obvious reason why this should have been true of the ancestral *Synthliboramphus* rather than of other auks, but as no information is available on the survival of any of them during the immediate post-departure period, either then or now, it is hard to completely eliminate the possibility.

HOW DO THESE IDEAS FIT THE EVIDENCE FOR ANCIENT MURRELETS?

The idea that precocial departure in *Synthliboramphus* has evolved because of constraints on the ability of the parents to provision the chicks seems to be a logical extension of the process that brought about the evolution of the intermediate strategy. However, some features of the murrelets' biology make this seem unlikely. Firstly, as relatively small alcids, they should be capable of carrying proportionately more food to their chicks than the intermediate species. Their weight to wing-area ratio (1.02 g/cm^2) is considerably lower than those found in intermediate species [Razorbill, 1.63; Common Murre, 2.06; Thick-billed Murre, 1.69 (Livezey, 1988)]. At Reef Island, we often observed that Ancient Murrelets were capable of taking off vertically, without wind-assistance. This ability suggests that they have a much better potential load-carrying capacity than guillemots, which have to taxi along the surface for some distance before getting airborne. If the load-carrying capacity of the murrelets is similar to the other small auks, then any constraint on their ability to provision chicks must derive either from the need to forage over exceptionally large distances, which is not supported by the evidence presented in Chapter 3, or from some peculiarity of the food supply, such as the unpredictability postulated by Sealy.

Other small, plankton-feeding auks develop pouches in their throats during the breeding season, which allow them to store many small food items for delivery to their chicks (Speich and Manuwal, 1974). *Synthliboramphus* murrelets do not possess such pouches. Whether their ancestors had pouches, or whether they fed their chicks on larger fish carried in the bill, like the *Brachyramphus* murrelets, we have no way of knowing. If *Synthliboramphus* did develop pouches at one time, the evolution of precocial chicks would have made them redundant. As they develop and regress annually in the auklets, there would probably be little difficulty in selection operating to eliminate them. In any case, if we assume that *Synthliboramphus* evolved from a semi-precocial ancestor that laid two eggs, we have to assume that the ancestor could carry enough food to provision two chicks. We cannot use the current inability of *Synthliboramphus* murrelets to carry large loads of their normal food as an explanation for their chicks' precocity (as Bedard, 1969a, attempted); cause and effect must work in the other direction.

Sealy's hypothesis, that precocity evolved because of unpredictable feeding conditions, has been accepted by others (e.g. Birkhead and Harris, 1985). However, apart from Sealy's own, purely qualitative observations, there is little information on whether the food of Ancient Murrelets is especially unpredictable. Over much of their range they feed on euphausids and young sandlance during the breeding season (Chapter 5), and these aggregate in large swarms, which may be very patchy. However, they are also important prey for other seabirds; Black-legged Kittiwakes, Cassin's Auklets (Sanger, 1987; Burger and Powell, 1990), which do not have precocial chicks. That does not eliminate the possibility that the patchiness of their prey has been a major selection pressure in the evolution of precocity, because many organisms have evolved different solutions to similar ecological predicaments.

There is, however, a logical objection to Sealy's theory that I find quite compelling. If Ancient Murrelets found it hard to locate concentrations of prey to feed chicks at the nest, when they were free to range as far as they could fly, they would surely have even more difficulty once the chicks were at sea. At that time they would be confined to searching areas within swimming range. If they flew off in search of food they would run the risk of not being able to locate their chicks again. Taking the young to the food source suggests *a priori* that the food supply is predictable, rather than otherwise. Those seabirds which feed on very unpredictable food sources, such as albatrosses, tend to be among the most peripatetic species (Weimerskirch et al., 1988). Our observations in Hecate Strait, admittedly involving only a small sample, suggested that family parties of Ancient Murrelets spread out once they had left the island, but fed within one or at most two hours' flying time of the colony (Duncan and Gaston, 1990). Puffins, Brunnich's Guillemots and Dovekies all spend this sort of time travelling to and fro to provision their chicks (Bradstreet and Brown, 1985).

There is one other logical flaw in any theory which accounts for precocity in murrelets solely on the basis of feeding constraints. This is that, in all other offshore-feeding seabirds, the constraint on provisioning imposed by distant foraging has led to the evolution of a single-egg clutch. If the murrelets evolved precocial young because they could not feed them adequately in the nest, it is hard to see why, like the guillemots and the Razorbill, they did not take the initial step of reducing their clutch size, and hence their brood size, to one. This argument suggests that, although constraints on the ability of parents to provision their chicks may be a necessary component of any explanation for precocity in murrelets, it is not sufficient on its own.

The possibility that precocity has evolved in response to a high rate of predation of chicks on the colony is hard to evaluate, mainly because we know nothing about how they survive once they reach the water. However, it is clear that, over most of their range, chicks would be vulnerable, especially while small, to predation by deer mice or voles. I have argued earlier (Chapter 11) that on Reef Island, adult murrelets can successfully repel deer mice. Otherwise, it is hard to explain why only chicks, or pipping eggs, deserted by their parents are eaten. Similarly, predation by mice on Xantus' Murrelet eggs left unattended is high (Murray et al., 1983), and would probably be equally high on chicks if the parents were not present. Hence, if *Synthliboramphus* chicks were reared in the nest on islands where there were mice, one parent would need to remain in the burrow at all times during the nestling period, or at least until the chick was well enough grown to defend itself, meaning that only one parent could forage at a time. This is the case for *Uria* species, which have to defend their chicks on open ledges against gulls and ravens (Harris and Birkhead, 1985). The fact that only one parent at a time can forage halves the potential rate at which the chick can be fed, and may well have contributed to the evolution of the intermediate departure strategy. However, the murrelets presumably always had the option of breeding on islands without mice, as storm petrels, which have semi-precocial chicks, have to do.

Ancient Murrelets take great pains to keep their burrows free of odours, and other signs of occupation. It is hard not to conclude that the attentions of scent-hunting, mammalian predators must have had something to do with the evolution of such concealment behaviour. Alcid chicks reared in burrows naturally have to defecate in them, so that their burrows rapidly develop a strong smell, compounded by the decomposition of any food remains that are accidentally dropped. Such burrows would be impossible to conceal from a mammalian predator. Hence, whatever selection pressures led to the evolution of burrow concealment behaviour in Ancient Murrelets would presumably also favour a reduction in the length of the nestling period.

SOME SPECULATIONS ON THE ORIGINS OF PRECOCIALITY

Theories about life-history strategies suggest that different demographic parameters should be interrelated (Williams, 1966; Goodman, 1974; Stearns, 1976). Promislow and Harvey (1990) showed that, once weight has been taken into account, mortality is the best predictor of life-history strategy in mammals. A trade-off between the number of eggs laid and the survival of adults has been demonstrated for North American game birds by Zammuto (1986), and for all birds by Gaillard *et al.* (1989). If the Ancient Murrelet differs from other alcids in demographic characteristics other than clutch size, then these demographic differences may have contributed to the evolution of its peculiar life-history strategy.

If we accept that the ancestors of *Synthliboramphus* reared two or more semi-precocial chicks, then we must presume that they fed close to their breeding sites, like the modern *Cepphus* guillemots. At some stage *Synthliboramphus* changed to foraging offshore, but unlike other seabird genera it retained the ability to rear two young. The transition to offshore feeding must either have followed the evolution of precocity or developed at the same time. Retention of the two-egg clutch suggests that the demography of the ancestral population was such that adult survival was too low to permit the reduction in reproductive rate inherent in the adoption of a one-egg clutch.

The estimated annual survival of breeding Ancient Murrelets, at 75%, is lower than has been observed for any other alcid or any other pelagic seabird (Hudson, 1985; Croxall and Gaston, 1988). The lowest annual adult survival rate otherwise for an alcid is 81% (Manuwal, 1974), recorded for Cassin's Auklet at the Farallon Islands. Although survival rates are not available for the other *Synthliboramphus* species, both Xantus' and Japanese Murrelets are known to suffer heavy mortality on their breeding colonies (Higuchi, 1979; Murray *et al.*, 1983).

In order to survive in the face of this high mortality, a high reproductive rate is needed, and this requires the potential to rear more than one offspring per year. The normal annual productivity to departure, about 1.5 chicks per breeding pair, is the highest regularly recorded for any auk, although similar productivity has been recorded for a few populations of Black Guillemots (Harris and Birkhead, 1985). As an offshore feeder, and nocturnal colony visitor, the Ancient Murrelet can only acheive this reproductive output by taking the chicks to the feeding area, a practice which also eliminates the risk of predation during provisioning and lowers parental mortality. Hence, I propose that the need to maintain a high reproductive rate and to reduce the vulnerability of breeders to mortality while visiting the colony, contributed to the evolution of precocity in *Synthliboramphus*. I recognize that this argument depends on the assumption that the mortality of chicks after leaving the colony is not markedly greater than the mortality suffered by other young auks during the pre-fledging period. The vulnerability of adult murrelets while visiting the colony may well be a major cause of the low adult survival rate.

As I indicated at the outset, no single chain of cause and effect can be designed to account for precocity in murrelets. A high risk of predation while visiting the colony would have caused selection for traits that reduced the number of visits made by parents to the breeding site. This could have been acheived by reducing the brood size, but only at the price of reducing the annual reproductive output of the pair.

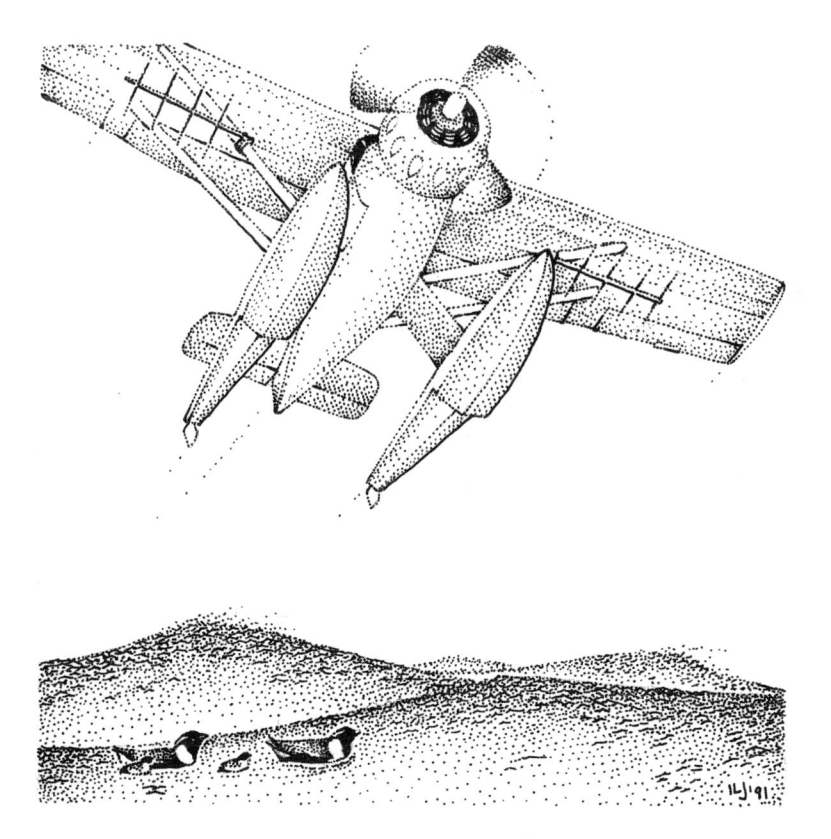

Two other factors may have been important in the evolution of precocity. Firstly, the development of precocity required many adaptations by the chicks, among which is the ability to thermoregulate while living on water. Craveri's Murrelet lives in warmer water than any other auk, and all four species of *Synthliboramphus* inhabit warmer waters than the majority of their family. It may be that their ancestors occupied relatively warm waters compared to other auks, and that this facilitated the transition to precocity. The expansion of the genus into the temperate and subarctic areas currently occupied by the Ancient Murrelet may have occurred after the evolution of precocity. Secondly, their generally southerly distribution enables them to take advantage of nocturnality, a strategy not available to arctic-breeding auks. It is doubtful whether the passage of chicks from their burrows to the sea could be achieved without the cover of darkness, because the chicks would be very vulnerable to predation by gulls, crows and other birds.

CHAPTER 16

Conservation

Why we do this type of research and what we have learned that may be of relevence to protecting Ancient Murrelets in the Queen Charlotte Islands.

SOME THOUGHTS ON CONSERVATION BIOLOGY

While we were working on Reef Island, we used to have occasional visits from tourists, who were travelling on the several large yachts that cruise the Queen Charlotte Islands each summer. I used to talk to them about the work that we were doing, and how it fitted into the larger framework of conservation in Canada. On these occasions, I was often asked about the status of the Ancient Murrelet, and whether it was declining. I generally answered that, although populations had declined precipitously in the Aleutians and on Langara Island, on most islands in the Queen Charlottes, the birds seemed to be holding their own. While people were generally pleased to hear that, they sometimes expressed surprise that the government was spending money on a species which, apparently, was not immediately endangered.

It is perfectly true that, if we read the papers and watch the television, we are daily reminded that this planet is being savagely denuded of its natural ecosystems. With catastrophe all around us, it seems like a luxury to study something not in immediate peril. However, if we spend all our time and energy on the most critically endangered species and ecosystems, we run several risks.

Some problems that we have set in train on this planet are only reversible at enormous cost. The buffalo will never again dominate the grasslands of the prairies; the spouts of right whales will never again grace the Bay of Biscay, or tigers stalk beside the walls of Agra. I shall not see walruses in the Gulf of St Lawrence, or wolves in the Scottish glens, in my lifetime. Likewise, there are probably many species that are bound to become extinct soon, irrespective of our efforts. Money put towards saving such species as the California Condor, especially where their natural habitat has been completely destroyed, may end in a bottomless pit. Moreover, even if they survive, these species may do so only in reserves that amount to no more than glorified zoos.

If we are going to wait until a species is endangered before taking an interest in it, we shall always be trying to catch up. We shall know little about the functioning of healthy populations, and a lot about small, possibly unrepresentative, fragments. Moreover, an endangered population is not a satisfactory one on which to conduct many types of research. By the time it has been reduced to a few hundred individuals, it is hard to justify any form of intrusion. Yet banding, or other forms of marking, nearly always involve capturing birds, which can hardly be done without disturbance.

To understand the factors that control natural populations generally requires many years of research. No amount of money can buy information in one or two years that requires many years to collect. This is a major weakness of much environmental assessment work. Because the Ancient Murrelet gets through its life-cycle relatively quickly, compared with other seabirds, I have been able to write this book based on only six years of work. However, many of our findings are preliminary, and it will take many more years before some aspects of inter-year variation, and inter-island movement, become clear.

In my opinion, our efforts are best directed at helping to conserve ecosystems that bear at least a passing resemblance to those which might have occurred naturally, before people with fire and machines and gunpowder reached them. The marine and terrestrial ecosystems of the Queen Charlotte Islands, though not untouched, surely qualify for this treatment. The Ancient Murrelet is an abundant bird in the archipelago, forming an important component of the ecosystem on many of the smaller islands, and having an interesting biology, differing in many respects from those of other marine birds. Its present world population is much diminished compared with what it was in the past. Nevertheless, some populations are large, and may live in conditions that approximate those under which the species evolved its distinctive adaptations. Understanding its biology helps us to understand the functioning of the ecosystems of which it forms a part. In addition, the information that we collected has many implications for monitoring Ancient Murrelet populations, and for conserving the species.

CONSERVATION OF ANCIENT MURRELETS

Ancient Murrelets are better able than most seabirds to withstand persistent mortality from predators, or chronic pollution. They have to be, because, as we have seen, they usually do not live nearly as long as most seabirds. They balance this by breeding at a relatively young age, and producing more young per year than most auks. Their greater reproductive output gives them the ability to recover faster from an environmental catastrophe than most marine birds.

The tendency of Ancient Murrelets to shift their burrow locations within a colony over a period of years, and the likelihood that young birds may recruit as breeders at any one of a number of colonies, not just the one where they were reared, may allow murrelet populations to relocate fairly rapidly in the face of disturbance to their breeding sites. On the other hand, Ancient Murrelets are extremely vulnerable to disturbance of their burrows, even the slightest intrusion carrying the risk of desertion. Their behaviour also leaves them very susceptible to predation by mammals, such as raccoons, foxes and otters.

The Queen Charlotte Islands have been subjected to outside influences for only two centuries. Although the industrial and agricultural worlds have had a great impact on their ecosystems, and although the formerly enormous seabird colony at Langara Island has been decimated, many of the smaller offshore islands have never been logged, and probably retain a vegetation and fauna similar to those that they supported before the Europeans arrived. The simplicity of their terrestrial ecosystems means that, in the Queen Charlottes, even small islands support a major fraction of the archipelago's complement of species. Hence, unlike small islands in some parts of the world, they preserve a reasonable representation of pre-existing ecosystems.

The main threats to the continued maintenance of these island ecosystems come from introduced mammals and accidental fires. Both of these hazards need to be carefully guarded against in the current situation, with tourism rapidly expanding in the Queen Charlotte islands. In view of all the other abuses which the archipelago has suffered, it seems the greatest good fortune that mink were never successfully established in the islands, and that cats, which must surely have been present in many of the South Moresby settlements, did not survive the withdrawal of their human sponsors. At present rats and raccoons pose the greatest threats to the islands' fauna, and one that only resolute action will contain.

The Ancient Murrelets remain vulnerable to changes in the marine environment which might affect them directly, or reduce their food supplies. The discovery of oil offshore in Hecate Strait would pose a potential threat requiring careful monitoring. With what we have learnt about the murrelets, we are in a good position to monitor any effects that developments might create. By counting the numbers of chicks departing through our funnels we can readily assess changes in breeding populations and in the timing of laying, without the disturbance involved in combing the colony to count burrows and inspect their contents. By banding and retrapping adult birds in their burrows, after chicks have hatched, we also may be able to detect changes in adult survival rates. These are critical data in determining whether the population is being affected by pollution or disturbance. However, it requires years of routine, repetitive observations to provide the right type of information.

In 1990, the task of monitoring Ancient Murrelet populations was taken over by the Laskeek Bay Conservation Society, a group of local people committed to advancing conservation and environmental education. They intend to continue trapping chicks and adult Ancient Murrelets on colonies in the South Moresby area. This programme is intended to assess changes in populations, and also to provide a situation in which interested people, whether residents or visitors, can obtain an experience of the remarkable world of the Ancient Murrelet. Surely, the more people that are familiar with the lives of these fascinating and endearing birds, the less they will countenance developments that may threaten them. It is one thing for government agencies to make plans and regulations, but without a broad base of support among the general public, such policies may be of little value. I wish the Laskeek Bay Conservation Society every success in their undertaking.

EPILOGUE

It is evident that Ancient Murrelets have a very different opinion of the sea from that of the Ud'din Attar's Hoopoe. To the Hoopoe the sea was a fickle, untrustworthy environment, full of lurking dangers. It holds dangers enough for the Ancient Murrelet, too, but they seem small compared to those provided by the land. For the murrelets, every visit to land carries risk; they must come and go under cover of dark, by day they cower in their burrows, and then only when incubation demands. As soon as the chicks are able, they are conveyed to the security of ocean, out of sight of land.

We may consider the albatrosses, those peerless travellers of the great oceans, or the penguins, which have forsaken flight entirely for the greater joys of swimming, as the most marine of birds. For the great antipathy that they show towards land, though, we must surely rank the murrelets with them, as showing extreme adaptations to the marine environment.

RHYME OF THE ANCIENT MURRELET

To the deep forests of the night they come
From far at sea, among the storm-tossed waves
Where spruce and hemlock rear their spreading crowns
And twining roots enclose their tiny caves

Where light and sun may never pierce the gloom
In shaft and passage far beneath the ground
Among the roots they guard their precious hoard
In chambers dark their treasure can be found

When dusk has passed and clouds obscure the stars
Their songs extol the dying of the light
Turning the silence of the evening calm
Into the shrill cacophony of night

Their calling done, each to its burrow creeps
And serenades its partner underground
Then before dawn's light strikes the shore
The minstrels flee far out upon the sound

AJG, 1984

APPENDIX 1

Scientific names of animals and plants mentioned in the text

BIRDS
Gaviiformes
Gaviidae
   Pacific Loon *Gavia pacifica*

Procellariiformes
Procellariidae
   Sooty Shearwater *Puffinus griseus*
Hydrobatidae
   Fork-tailed Storm Petrel *Oceanodroma furcata*
   Leach's Storm Petrel *Oceanodroma leucorhoa*

Pelecaniformes
Phalacrocoracidae
   Double-crested Cormorant *Phalacrocorax auritus*
   Pelagic Cormorant *Phalacrocorax pelagicus*

Anseriformes
Anatidae
   Canada Goose *Branta canadensis*
   Common Eider *Somateria mollissima*
   Oldsquaw *Clangula hyemalis*

Falconiformes
Accipitridae
   Bald Eagle *Halaeetus leucocephalus*
   Sharp-shinned Hawk *Accipiter striatus*
Falconidae
   Peregrine Falcon *Falco peregrinus pealei*

Charadriiformes
Haematopodidae
   Black Oystercatcher *Haematopus bachmani*
Scolopacidae
   Red-necked Phalarope *Phalaropus lobatus*
Laridae
   Herring Gull *Larus argentatus*
   Glaucous-winged Gull *Larus glaucescens*
   Black-legged Kittiwake *Rissa tridactyla*

Alcidae
   Dovekie (Little Auk) *Alle alle*
   Thick-billed Murre (Brunnich's Guillemot) *Uria lomvia*
   Common Murre (Guillemot) *Uria aalge*
   Razorbill *Alca torda*
   Black Guillemot *Cepphus grylle*
   Pigeon Guillemot *Cepphus columba*
   Marbled Murrelet *Brachyramphus marmoratus*
   Kittlitz' Murrelet *Brachyramphus brevirostris*
   Xantus' Murrelet *Synthliboramphus hypoleuca*
   Craveri's Murrelet *Synthliboramphus craveri*
   Ancient Murrelet *Synthliboramphus antiquus*
   Japanese (Crested) Murrelet *Synthliboramphus wumizusume*
   Cassin's Auklet *Ptychoramphus aleuticus*
   Whiskered Auklet *Aethia pusilla*
   Parrakeet Auklet *Cyclorhynchus psittacula*
   Crested Auklet *Aethia cristatella*
   Rhinoceros Auklet *Cerorhinca monocerata*
   Atlantic Puffin *Fratercula arctica*
   Horned Puffin *Fratercula corniculata*
   Tufted Puffin *Lunda cirrhata*

Strigiformes
Strigidae
   Northern Saw-whet Owl *Aegolius acadicus*

Apodiformes
Trochilidae
   Rufous Hummingbird *Selasophorus rufus*

Piciformes
Picidae
   Red-breasted Sapsucker *Sphyrapicus ruber*
   Hairy Woodpecker *Dryocopus villosus*
   Northern Flicker *Colaptes auratus*

Passeriformes
Corvidae
   Northwestern Crow *Corvus caurinus*
   Common Raven *Corvus corax*
   Steller's Jay *Cyanocitta stelleri*
Paridae
   Chestnut-backed Chickadee *Parus rufescens*
Certhidae
   Brown Creeper *Certhia familiaris*
Troglodilidae
   Winter Wren *Troglodytes troglodytes*
Muscicapidae
   Golden-crowned Kinglet *Regulus satrapa*
   Varied Thrush *Ixoreus naevius*
   Hermit Thrush *Catharus guttata*
   Swainson's Thrush *Catharus ustulatus*
Emberizidae
   Orange-crowned Warbler *Vermivora celata*
   Townsend's Warbler *Dendroica townsendi*
   Fox Sparrow *Passerella iliaca*
   Song Sparrow *Melospiza melodia*
Fringillidae
   Pine Siskin *Carduelis pinus*
   Red Crossbill *Loxia curvirostra*

MAMMALS
Little Brown Bat *Myotis lucifugus*
Shrew *Sorex monticolus*
Dawson Caribou *Rangifer tarandus dawsoni*
Sitka Black-tailed Deer *Odocoileus hemionus sitkensis*
Rocky Mountain Elk *Cervus elephas nelsoni*
Raccoon *Procyon lotor*
Pine Marten *Martes americana*
Ermine *Mustela erminea*
American Mink *Mustela vison*
River Otter *Lutra canadensis*
Sea Otter *Enhydra lutris*
Black Bear *Ursus americanus carlottae*
Grizzly Bear *Ursus arctos*
Beaver *Castor canadensis*
Muskrat *Ondatra zibethicus*
Deer Mouse *Peromyscus maniculatus* and *Peromyscus sitkensis*
Ship Rat *Rattus rattus*
Harbour Seal *Phoca vitulina*

Northern Elephant Seal *Mirounga angustirostris*
Steller's (Northern) Sea Lion *Eumetopias jubata*
Northern Fur Seal *Callorhinus ursinus*
Humpback Whale *Megaptera novaeangliae*
Fin Whale *Balaenoptera physalus*
Grey Whale *Eschrichtius robustus*
Killer Whale *Orcinus orca*
Dall's Porpoise *Phocoenoides dalli*
Pacific White-sided Dolphin *Lagenorhynchus obliquidens*

FISH AND MARINE INVERTEBRATES
Purple Sea-urchin *Strongylocentrotus purpuratus*
Cutthroat Trout *Salmo clarki*
Steelhead *Salmo gairdneri*
Salmon *Onchorynchus* spp.
Halibut *Hippoglossus stenolepis*
Herring *Clupea harengus*
Capelin *Mallotus villosus*
Walleye Pollock *Theragra chalcogramma*
Pacific sandlance *Ammodytes hexapterus*
Seaperch *Cymatogaster aggregata*
Rockfish *Sebastes* spp.

PLANTS
Western red cedar *Thuja plicata*
Western hemlock *Tsuga heterophylla*
Sitka spruce *Picea sitchensis*
Yellow Cedar *Chamaecyparis nootkatensis*
Yew *Taxus brevifolia*
Lodgepole pine *Pinus contorta*
Alder *Alnus rubra, A. sinuata*
Holly *Ilex* sp.
Single delight *Moneses uniflora*
Indian pipes *Monotropa uniflora*
Skunk Cabbage *Lysichitum americanum*
Huckleberry *Vaccinium parvifolium*
Salal *Gaulteria shallon*
Salmonberry *Rubus spectabilis*
Potentilla *Potentilla villosa*
Red Columbine *Aquilegia formosa*
Monkeyflower *Mimulus guttatus*
Lupins *Lupinus* sp.
Nettles *Urtica* sp.
Chocolate lily *Fritillaria camschatcensis*
Calypso Orchid *Calypso bulbosa*

APPENDIX 2

Census details for Reef Island, 1989

Transect No.	Date	Distance to: Plot 1	Distance to: End	Plot No. 1	2	3	4	5	6	7	8	Total burrows
W12	5/6	Outside colony										
W11	5/6	5	280	0	0	0	4	0	0			4
W10	5/6	30	350	0	0	2	0	0	0	0	0	2
W09	19/5	5	380	1	1	7	0	0	1	0	0	10
W08	1/5	30	150	1	5	0						6
W07	1/5	5	130	0	3	0						3
W06	30/4	30	155	5	6	3	0					14
W05	30/4	5	180	0	0	0	0					0
W04	29/4	30	255	1	0	0	1	0				2
W03*	29/4	5	230	4	1	3	7	0				15
W02*	28/4	30	350	0	0	0	1	0	0	0		1
E01*	12/4	5	330	0	0	3	5	5	1	4		18
E02*	18/4	30	300	0	0	0	3	0	3	0		6
E03*	18/4	5	230	0	2	2	2	0				6
E04*	19/4	30	205	0	2	0	0					2
E05*	19/4	5	130	0	1	1						2
E06*	21/4	30	155	0	6	0						6
E07*	21/4	5	130	1	0	4						5
E08	21/4	30	155	1	0	0						1
E09	23/4	5	130	2	1	1						4
E10	23/4	30	155	7	8	1						16
E11	24/4	5	225	2	1	3	2	4				12
E12	24/4	30	155	1	1	2						4
E13	25/4	5	80	0	0							0
E14	28/4	30	55	1								1
E15	28/4	Outside colony										
S01	20/5	30	350	0	0	0	0	1	5	0		6
S02	25/5	5	230	0	0	0	0	2				2
S03	5/6	5	130	1	1	0						2
S04	5/6	5	230	0	0	0	0	0				0

* Transects spaced at 50 m interval.

References

Ainley, D.G. (1976). The occurrence of seabirds in the coastal region of California. *Western Birds*, 7: 33–68.
Ainley, D.G. and Boekelheide, R.J. (1990). *Seabirds of the Farallon Islands*. Stanford University Press, Stanford. 450 pp.
Ainley, D.G., Degange, A.R., Jones, L.L. and Beach, R.J. (1981). Mortality of seabirds in high seas salmon gill nets. *Fishery Bulletin*, 79: 800–806.
Anderson, D.W. and Hickey, J.J. (1972). Eggshell changes in certain North American birds. *Proceedings of the International Ornithological Congress*, 15: 514–540.
American Ornithologists' Union (AOU). (1886). *Code of Nomenclature and Check-list of North American Birds*. American Ornithologists' Union, New York.
American Ornithologists' Union (AOU). (1910). *Check-list of North American Birds*, 3rd ed. American Ornithologists' Union, New York.
American Ornithologists' Union (AOU). (1983). *Check-list of North American Birds*, 6th ed. American Ornithologists' Union, New York.
Anon. (1936). *Annual Report of the British Columbia Game Commission, 1936*. Government of British Columbia, Victoria.
Anon. (1974). *Atlas of the Oceans—Pacific Ocean*. Ministerstvo Oborony SSSR. Voenno-Morskoi Flot.
Arnold, L.W. (1948). Observations on populations of North Pacific pelagic birds. *Auk*, 65: 553–558.
Asbirk, S. (1979). The adaptive significance of the reproductive pattern in the Black Guillemot *Cepphus grylle*. *Videnskabelige Meddelelser Dansk Naturhistorisk Forening*, 141: 29–80.
Ashmole, N.P. (1963). The regulation of numbers of tropical oceanic birds. *Ibis*, 103b: 458–473.
Ashmole, N.P. (1971). Seabird ecology and the marine environment. In *Avian Biology, I*, Farner, D.S. and King, J.R. (eds), pp. 224–286. Academic Press, New York.
Astheimer, L.B. and Grau, C.R. (1990). A comparison of yolk growth rates in seabird eggs. *Ibis*, 132: 380–394.
Atkinson, I.A.E. (1985). The spread of commensal species of *Rattus* to oceanic islands and their effects on island avifaunas. In *Conservation of Island Birds*, Moors, P.J. (ed.) pp. 35–84. International Council for Bird Preservation, Cambridge.
Attar, Farid Ud-din. (1984). *The Conference of the Birds*. Trans. by Darbandi, A. and Davis, D. Penguin, Harmondsworth, U.K.
Audubon, J.J. (1827–1830). *The Birds of America*. Published by the Author (reproduced by Macmillan, New York, 1937).
Audubon, J.J. (1839). *Ornithological Biographies* V: 100.
Austin, O.L. (1948). The birds of Korea. *Bulletin of the Museum of Comparative Zoology, Harvard*, Vol 101: No. 1, 301 pp.
Austin, O.L. and Kuroda, N. (1953). The birds of Japan, their status and distribution. *Bulletin of the Museum of Comparative Zoology, Harvard*, 109: 280–637.
Bailey, E.P. (1978). Breeding seabird distribution and abundance in the Shumagin Islands, Alaska. *Murrelet*, 59: 82–91.
Bailey, E.P. and Faust, N.H. (1980). Summer distribution and abundance of marine birds and mammals in the Sandman Reefs, Alaska. *Murrelet*, 61: 6–19.

Bailey, E.P. and Faust, N.H. (1981). Summer distribution and abundance of marine birds and mammals between Mitrofania and Sutwik Islands South of the Alaska Peninsula. *Murrelet*, 62: 34–42.
Bailey, E.P. and Faust, N.H. (1984). Distribution and abundance of marine birds breeding between Amber and Kamishak Bays, Alaska, with notes on interactions with bears. *Western Birds* 15: 161–174.
Bailey, E.P. and Kaiser, G.W. (in press). Impacts of introduced predators on nesting seabirds in the Northeast Pacific. In *Proceedings of a Symposium on the Status of Seabirds in the North Pacific, February 1990*, Vermeer, K (ed.). Canadian Wildlife Service, Ottawa.
Bailey, E.P. and Trapp, J.L. (1986). A reconnaissance of breeding marine birds and mammals in the East-Central Aleutian Islands—Kasatochi to the Islands of Four Mountains—summer 1982, with notes on other species. *U.S. Fish and Wildlife Service Ms. Report*. Anchorage, Alaska.
Baird, S.F., Brewer, T.M. and Ridgeway, R. (1884). *The Water Birds of North America*. Little, Brown and Co., Boston.
Balmer, A. (1936). Beach combings at Westport, Washington. *Murrelet*, 17: 21.
Balz, D.M. and Morejohn, G.V. (1977). Food habits and niche overlap of seabirds wintering on Monterey Bay, California. *Auk*, 94: 526–543.
Bartonek, J.C. and Gibson, D.D. (1972). Summer distribution of pelagic birds in Bristol Bay, Alaska. *Condor*, 74: 416–422.
Bedard, J. (1966). New records of Alcids from St Lawrence Island, Alaska. *Condor*, 68: 503–506.
Bedard, J. (1969a). Adaptive radiation in Alcidae. *Ibis*, 111: 189–198.
Bedard, J. (1969b). Feeding of the Least, Crested and Parrakeet Auklets on St. Lawrence Island, Alaska. *Canadian Journal of Zoology*, 47: 1025–1050.
Bedard, J. (1969c). Nesting of the Least, Crested and Parrakeet auklets on St. Lawrence Island, Alaska. *Condor*, 71: 386–398.
Beebe, F.L. (1960). The marine peregrines of the northwest Pacific coast. *Condor*, 62: 145–189.
Bendire, C. (1895). Notes on the Ancient Murrelet (*Synthliboramphus antiquus*), by Chase Littlejohn. With annotations. *Auk*, 12: 270–278.
Bengtson, S-A. (1984). Breeding ecology and extinction of the Great Auk (*Pinguinus impennis*): anecdotal evidence and conjectures. *Auk*, 101:1–10.
Bent, A.C. (1919). Life histories of North American diving birds. *Smithsonian Institute Bulletin* No. 107.
Bertram, D.F. (1989). The status of Ancient Murrelets breeding on Langara Island, British Columbia, in 1988. *Canadian Wildlife Service Technical Report Series* No. 59, Canadian Wildlife Service, Pacific and Yukon Region, Delta, British Columbia.
Bertram, D.F. and Kaiser, G.W. (1988). Monitoring growth and diets of nestling Rhinoceros Auklets to guage prey availability. *Canadian Wildlife Service Technical Report Series* No. 48, Canadian Wildlife Service, Pacific and Yukon Region, Delta, British Columbia.
Bigg, M.A. (1989). Status of the Steller Sea Lion, *Eumetopias jubatus*, in Canada. *Canadian Field-Naturalist*, 102: 315–336.
Birkhead, T.R. (1977). Adaptive significance of the nestling period of Guillemots *Uria aalge*. *Ibis*, 119: 544–549.
Birkhead, T.R. (1985). Coloniality and social behaviour in the Atlantic Alcidae. In *The Atlantic Alcidae*, Nettleship, D.N. and Birkhead, T.R. (eds), pp. 355–383. Academic Press, London.
Birkhead, T.R. and Gaston, A.J. (1988). The composition of Ancient Murrelet eggs. *Condor*, 90: 965–966.
Birkhead, T.R. and Harris, M.P. (1985). Ecological adaptations for breeding in the

Atlantic Alcidae. In *The Atlantic Alcidae*, Nettleship, D.N. and Birkhead, T.R. (eds), pp. 205–232. Academic Press, London.
Birkhead, T.R. and Nettleship, D.N. (1982). The adaptive significance of egg size and laying date in Thick-billed Murres *Uria lomvia*. *Ecology*, 63: 300–306.
Blakiston, T. and Pryer, H. (1878). A catalogue of the birds of Japan. *Ibis 4th series*, 7: 209–250.
Boersma, P.D. and Wheelwright, N.T. (1979). Egg neglect in the Procellariiformes: reproductive adaptations in the Fork-tailed Storm-Petrel. *Condor*, 81: 157–165.
Bonaparte, C.J. (1838). *A Geographical and Comparative List of the Birds of Europe and North America*. London.
Bradstreet, M.S.W. and Brown, R.G.B. (1985). Feeding ecology of the Atlantic Alcidae. In *The Atlantic Alcidae*, Nettleship, D.N. and Birkhead, T.R. (eds), pp. 262–318. Academic Press, London.
Brandt, J.F. (1837). *Bulletin of the Academy of St. Petersburg*, II, 347.
Brandt, J.F. (1869). Erganzungen und Berichtigungen zur Naturgeschichte der Familie der Alciden. *Melanges Biologiques, Bulletin of the Imperial Academy of Science, St. Petersburg*, 7: 199–268.
Brewer, T.M. (1840). *Wilson's American Ornithology*. Otis, Broaders and Co., Boston.
Brewster, W. (1902). Birds of the Cape region of Lower California. *Bulletin of the Museum of Comparative Zoology, Harvard*, 41.
Briggs, K.T., Tyler, W.B., Lewis, D.B. and Carlson, D.R. (1987). Bird communities at sea off California: 1975–1983. *Studies in Avian Biology*, 11.
Brooks, A. and Swarth, S.H. (1925). *A Distributional List of the Birds of British Columbia*. Pacific Coast Avifauna No. 17. Berkeley, California.
Brown, R. (1868). Synopsis of the birds of Vancouver Island. *Ibis IV*: 414–428.
Brown, R.G.B. (1985). The Atlantic Alcidae at sea. In *The Atlantic Alcidae*, Nettleship, D.N. and Birkhead, T.R. (eds), pp. 383–426. Academic Press, London.
Burger, A.E. (1991). Maximum diving depths and underwater foraging in alcids and penguins. In *Studies of High Latitude Seabirds. 1. Behavioural, Energetic and Oceanographic Aspects of Seabird Feeding Ecology*, Montevecchi, W.A. and Gaston, A.J. (eds), pp. 9–15. Canadian Wildlife Service Occasional Papers No. 68. C.W.S., Ottawa.
Burger, A.E. and Powell, D. (1990). Diving depths and diets of Cassin's Auklet at Reef Island, British Columbia. *Canadian Journal of Zoology*, 68: 1572–1577.
Busch, B.C. (1985). *The War Against the Seals*. McGill-Queen's University Press, Montreal.
Byrd, G.V. and Day, R.H. (1986). The avifauna of Buldir Island, Aleutian Islands, Alaska. *Arctic*, 39: 109–118.
Byrd, G.V., Johnson, D.L. and Gibson, D.D. (1974). The birds of Adak Island, Alaska. *Condor*, 76: 288–300.
Cairns, D.K. (1987). The ecology and energetics of chick provisioning by Black Guillemots. *Condor*, 89: 627–635.
Cameron, F.E. (1957). Some factors influencing the distribution of pelagic copepods in the Queen Charlotte Islands area. *Journal of the Fisheries Research Board of Canada*, 14: 165–202.
Campbell, R.W. (1968). Alexandrian rat predation on Ancient Murrelet eggs. *Murrelet*, 49: 38.
Campbell, R.W. and Garrioch, H.M. (1979). *Seabird Colonies of the Queen Charlotte Islands*. Map published by the Royal British Columbia Provincial Museum, Victoria.
Campbell, R.W., Dawe, N.K., McTaggart-Cowan, I., Cooper, J.M., Kaiser, G.W. and McNall, M.C.E. (1990). *The Birds of British Columbia*. Royal British Columbia Museum and the Canadian Wildlife Service, Victoria.
Carson, T. 1981. Birds that go thump in the night. *Wildlife Revue*, 9: 18–27.

Carey, C., Rahn, H. and Parisi, P. (1980). Calories, water, lipid and yolk in avian eggs. *Condor*, 82: 335–343.

Chalmers, M.G. (1986). *Annotated Checklist of the Birds of Hong Kong*, 4th ed. Hong Kong Bird Watching Society, Hong Kong.

Chang, J.W. (1980). *A Field Guide to the Birds of Taiwan*. Publisher unknown.

Chen Zhao-Qing. (1988). Niche selection of breeding seabirds on Chenlushan Island in the Yellow Sea, China. *Colonial Waterbirds*, 11: 306–307.

Cheng Tso-Hsin. (1987). *A Synopsis of the Avifauna of China*. Paul Parey Scientific Publishers, Hamburg and Berlin.

Childerhose, R.J. and Trim, M. (1979). *Pacific Salmon*. Douglas and McIntyre, Vancouver.

"C.L." (1789). *A Voyage round the World in the Years 1785, 1786, 1787 and 1788*. Reprinted 1984, Ye Galleon Press, Fairfield, Washington.

Clague, J.J. (1989). Quaternary geology of the Queen Charlotte Islands. In *The Outer Shores*, Scudder, G.G.E. and Gessler, N. (eds), pp. 65–74. Queen Charlotte Islands Museum Press, Skidegate, B.C.

Clobert, J. and Lebreton, J.D. (1985). Dependance de facteurs de milieu dans les estimations de taux de survie par capture-recapture. *Biometrics*, 41: 1031–1037.

Cody, M.L. (1973). Coexistence, coevolution and convergent evolution in seabird communities. *Ecology*, 54: 31–43.

Collar, N.J. and Andrew, P. (1988). *Birds to Watch: the ICBP World Checklist of Threatened Birds*. International Council for Bird Preservation Technical Publication No. 8. I.C.B.P., Cambridge.

Collins, B.T. and Gaston, A.J. (1987). Estimating the error involved in using egg density to predict laying dates. *Journal of Field Ornithology*, 58: 464–473.

Coues, E. (1868). *A Monograph of the Alcidae*. Proceedings of the Academy of Natural Sciences, Philadelphia.

Coues, E. (1882). *Check List of North American Birds*. Estes and Lauriat, Boston.

Coues, E. (1890). *Key to North American Birds, Revised Edition*. Estes and Lauriat, Boston.

Croxall, J.P. and Gaston, A.J. (1988). Patterns of reproduction in high-latitude northern- and southern-hemisphere seabirds. *Proceedings of the International Ornithological Congress*, 19: 1176–1194.

Cumming, R.A. (1931). Some birds observed in the Queen Charlotte Islands, British Columbia. *Murrelet*, 12: 15–17.

Dalzell, K.E. (1968). *The Queen Charlotte Islands*. Bill Ellis, Queen Charlotte City, B.C.

Darcus, S.J. (1930). Notes on birds of the northern part of the Queen Charlotte Islands in 1927. *Canadian Field Naturalist*, 44: 45–49.

Degange, A.R. (1983). Seabird mortality in the Japanese salmon mothership fishery summer 1982. *U.S. Fish and Wildlife Service Ms. Report*, Anchorage, Alaska.

Degange, A.R., Possardt, E.E. and Frazer, D.A. (1977). The breeding biology of seabirds on the Forrester Island National Wildlife Refuge, 15 May–1 September 1976. *U.S. Fish and Wildlife Service Ms. report*, Anchorage, Alaska.

Dement'ev, G.P. and Gladkov, N.A. (1969). *Birds of the Soviet Union*, Vol 2. Israel Programme for Scientific Translation, Jerusalem [published as 'Ptitsy Sovetskogo Soyuza', 1951].

Deweese, L.R. and Anderson, D.W. (1976). Distribution and breeding biology of Craveri's Murrelet. *Transactions of the San Diego Natural History Society*, 18: 155–168.

Dodimead, A.J. (1980). A general review of the oceanography of the Queen Charlotte Sound–Hecate Strait–Dixon Entrance region. *Canadian Ms. Report of Fisheries and Aquatic Sciences*, No. 1574.

Drent, R.H. and Guiget, C.J. (1961). A catalogue of British Columbia sea-bird colonies. *British Columbia Provincial Museum Occasional Paper*, No. 12. Victoria, British Columbia.

Duggins, D.O. (1983). Marine dominoes. *Equinox*, 2(7): 43–57

Duncan, D.C. and Gaston, A.J. (1988). The relationship between precocity and body composition in some neonate alcids. *Condor*, 90: 718–721.

Duncan, D.C. and Gaston, A.J. (1990). Movements of Ancient Murrelet broods away from a colony. *Studies in Avian Biology*, 14: 109–113.

Ebeling, A.W and Laur, D.R. (1987). Fish populations in kelp forests without sea otters: effects of severe storm damage and destructive sea urchin grazing. In *The Community Ecology of Sea Otters*. Vanblaricom, G.R. and Estes, J.A. (eds), pp. 169–191. Springer-Verlag, New York.

Elliot, J.E., Noble, D.G., Norstrom, R.J. and Whitehead, P.E. (1989). Organochlorine contaminants in seabird eggs from the Pacific coast of Canada, 1971–1986. *Environmental Monitoring and Assessment*, 12: 67–82.

Eppley, Z.A. (1984). Development of thermoregulatory abilities in Xantus' Murrelet chicks *Synthliboramphus hypoleucus*. *Physiological Zoology*, 57: 307–317.

Estes, J.A. and Vanblaricom, G.R. (1987). Concluding remarks. In *Community Ecology of Sea Otters*, Vanblaricom, G.R. and Estes, J.A. (eds), pp. 210–218. Springer-Verlag, New York.

Evans, P.G.H. (1987). *The Natural History of Whales and Dolphins*. Christopher Helm, London.

Fannin, J. (1891). *Check List of British Columbia Birds*. Publisher unknown.

Fleming, J.H. (1912). The Ancient Murrelet (*Synthliboramphus antiquus*) in Ontario. *Auk*, 29: 387–388.

Flint, V.E., Bohme, R.L., Kostin, Y.V. and Kuznetsov, A.A. (1984). *A Field Guide to the Birds of the USSR*. Princeton University Press, Princeton, New Jersey.

Flint, V.E. and Golovkin, A.N. (1990). Birds of the U.S.S.R.: Auks, murres and puffins (Alcidae). "Nauka", Moscow [Russian, translated into English by the Office of the Secretary of State, Government of Canada].

Ford, C. (1967). *Where the Sea Breaks its Back*. Victor Gollancz, London.

Forsell, D.J. and Gould, P.J. (1981). Distribution and abundance of marine birds and mammals wintering in the Kodiak area of Alaska. *U.S. Fish and Wildlife Service, Ms. Report*. Washington.

Freuchen, P. and Salomonsen, F. (1958). *The Arctic Year*. G.P.Putnam's Sons, New York.

Fujimaki, Y. (1986). Seabird colonies on Hokkaido Island. In *Morskie Ptitsy Dalnego Vostoka* [Seabirds of the Far East], Litvinenko, N.M. (ed.), pp. 82–87. Far East Science Centre, USSR Academy of Science, Institute of Biology and Soil Sciences, Vladivostok.

Furness, R.W. (1978). Movements and mortality rates of Great Skuas ringed in Scotland. *Bird Study*, 25: 229–238.

Gaillard, J.-M., Pontier, D., Allaine, D., Lebreton, J.-D., Trouvilliez, J. and Clobert, J. (1989). An analysis of demographic tactics in birds and mammals. *Oikos*, 56: 59–76.

Gaston, A.J. (1985). Development of the young in the Atlantic Alcidae. In *The Atlantic Alcidae*, Nettleship, D.N. and Birkhead, T.R. (eds), pp. 319–354. Academic Press, London.

Gaston, A.J. (1990). Population parameters of the Ancient Murrelet. *Condor*, 92: 998–1011.

Gaston, A.J. and Jones, I.L. (1989). The relative importance of stress and programmed anorexia in determining mass loss by incubating Ancient Murrelets. *Auk*, 106: 653–658.

Gaston, A.J. and Powell, D.W. (1989). Natural incubation, egg neglect, and hatchability in the Ancient Murrelet. *Auk*, 106: 433–438.

Gaston, A.J., Jones, I.L. and Noble, D.G. (1988a). Monitoring Ancient Murrelet breeding populations. *Colonial Waterbirds* 11: 58–66.

Gaston, A.J., Jones, I.L., Noble, D.G. and Smith, S.A. (1988b). Orientation of ancient murrelet, *Synthliboramphus antiquus*, chicks during their passage from the burrow to the sea. *Animal Behaviour*, 36: 300–303.

Gmelin, J.S. (1789). *Caroli a Linne Systema Naturae*, Tom I, Pars II, 554–555.

Goodman, D. (1974). Natural selection and a cost ceiling on reproductive effort. *American Naturalist*, 108: 247–268.

Gore, M.E.J. and Pyong-oh, W. (1971). *The Birds of Korea*. Royal Asiatic Society, Korea.

Gough, B.M. (1980). *Distant Dominion*. University of British Columbia Press, Vancouver, B.C.

Gould, P.J. (1977). Shipboard surveys of marine birds. In *Environmental Assessment of the Alaskan Continental Shelf, Annual Report of Principal Investigators for the Year Ending March 1977, vol. 3. Receptors—Birds*: pp. 192–284. MMOS, Anchorage, Alaska.

Green, C. de B. (1916). Notes on the distribution and nesting habits of *Falco peregrinus pealei* Ridgeway. *Ibis* (Ser 10), 4: 473–476.

Grinnell, J. and Miller, A.H. (1944). *The Distribution of the Birds of California*. Pacific Coast Avifauna, No 27.

Guenther, K. (1965). Ancient Murrelet near Spokane, Washington. *Murrelet*, 46: 11.

Guiguet, C.J. (1953). An ecological study of Goose Island, British Columbia, with special reference to mammals and birds. *Occasional Papers of the British Columbia Provincial Museum*, No. 10, Victoria.

Guiguet, C.J. (1972). *The Birds of British Columbia: (9) Diving Birds and Tube-nosed Swimmers*. British Columbia Provincial Museum Handbook #29, Victoria.

Hagelund, W.A. (1987). *Whalers no more*. Harbour Publishing, Madeira Park, B.C.

Harris, M.P. and Birkhead, T.R. (1985). Breeding ecology of the Atlantic Alcidae. In *The Atlantic Alcidae*, Nettleship, D.N. and Birkhead, T.R. (eds), pp. 155–204. Academic Press, London.

Harrison, C.S. (1977). Aerial surveys of marine birds. In *Environmental Assessment of the Alaskan Continental Shelf, Annual report of principal investigators for the year ending March 1977, vol. 3. Receptors—birds*. pp. 285–593.

Hasegawa, H. (1984). Status and conservation of seabirds in Japan, with special attention to the Short-tailed Albatross. In *Status and Conservation of the World's Seabirds*, Croxall, J.P., Evans, P.G.H. and Schreiber, R.W. (eds), pp. 487–500. ICBP Technical Publication No. 2, Cambridge.

Hatch, S.A. and Hatch, M.A. (1983). Populations and habitat use of marine birds in the Semidi Islands, Alaska. *Murrelet*, 64: 39–46.

Hatch, S.A., Byrd, G.V., Irons, D.B. and Hunt, G.L., Jr. (in press). Status and ecology of kittiwakes in the North Pacific. In *Proceedings of a Symposium on the Status of Seabirds in the North Pacific, February 1990*, K. Vermeer (ed.). Canadian Wildlife Service, Ottawa.

Heath, H. (1915). Birds observed on Forrester Island, Alaska, during the summer of 1913. *Condor*, 17: 20–41.

Higuchi, Y. (1979). Breeding ecology and distribution of the Japanese Crested Murrelet. *Aquatic Biology*, 1(3), 20–24 [in Japanese].

Hobson, K.A. (1991). Stable isotopic determinations of the trophic relationships of seabirds: preliminary investigations of alcids from coastal British Columbia. In *Studies of High-latitude Seabirds. 1. Behavioural, Energetic and Oceanographic Aspects of Seabird Feeding Ecology*, Montevecchi, W.A. and Gasto, , A.J. (eds), pp. 16–20. Canadian Wildlife Service Occasional Paper No. 68, Ottawa.

Hoffmann, R. (1924). Breeding of the Ancient Murrelet in Washington. *Condor*, 26: 191.

Hoffman, W., Heinemann, D. and Wiens, J.A. (1981). The ecology of seabird feeding flocks in Alaska. *Auk*, 98: 437–456.

Hora, B. (1981). *Trees of the World*. Oxford University Press, Oxford.

Howard, H. (1949). Avian fossils from the marine pleistocene of southern California. *Condor*, 51: 20–28.

Hudson, P.J. (1985). Population parameters for the Atlantic Alcidae. In *The Atlantic Alcidae*, Nettleship, D.N. and Birkhead, T.R. (eds), pp. 233–261. Academic Press, London.

Hunt, G.L., Jr., Pitman, R.L. and Jones, H.L. (1980). Distribution and abundance of seabirds breeding on the California Channel Islands. In *The California Islands: Proceedings of a Multi-disciplinary Symposium*, Powers, D. M. (ed.), pp. 443–459. Santa Barbara Museum of Natural History, Santa Barbara.

Ishizawa, T. (1933). Life history of *Synthliboramphus antiquus*. *Plants and Animals, Tokyo*, 1: 279–280 [in Japanese].

Islands Protection Society. (1984). *Islands at the Edge*. Douglas and McIntyre, Vancouver.

Isleib, M.E. and Kessel, M. (1973). Birds of the north Gulf coast—Prince William Sound region, Alaska. *Biological Papers of the University of Alaska*. 14, Fairbanks.

Jacob, F. (1982). *The Possible and the Actual*. Pantheon Books, New York.

Jehl, J.R., Jr and Bond, S.I. (1975). Morphological variation and species limits in murrelets of the genus *Endomychura*. *Transactions of the San Diego Natural History Society*, 18: 9–24.

Johanson, H. (1961). Revised list of the birds of the Commander Islands. *Auk*, 78: 44–56.

Johnstone, W.B. (1964). Two interior British Columbia records for the Ancient Murrelet. *Canadian Field-Naturalist*, 78: 199–200.

Jones, I.L. (1985). Structure and function of vocalizations and related behaviour of the Ancient Murrelet (*Synthliboramphus antiquus*). M.Sc. Thesis, University of Toronto.

Jones, I.L., Falls, J.B. and Gaston, A.J. (1987a). Vocal recognition betwen parents and young of ancient murrelets, *Synthliboramphus antiquus* (Aves: Alcidae). *Animal Behaviour*, 35: 1405–1415.

Jones, I.L., Falls, J.B. and Gaston, A.J. (1987b). Colony departure of family groups of Ancient Murrelets. *Condor*, 89: 940–943.

Jones, I.L., Falls, J.B. and Gaston, A.J. (1989). The vocal repertoire of the Ancient Murrelet. *Condor*, 91: 699–710.

Jones, I.L., Gaston, A.J. and Falls, J.B. (1990). Factors affecting colony attendance by Ancient Murrelets. *Canadian Journal of Zoology*, 68: 433–441.

Jones, R.D., Jr. and Byrd, G.V. (1979). Interrelationships between seabirds and introduced animals. In *Conservation of Marine Birds of Northern North America*, Bartonek, J.C. and Nettleship, D.N. (eds), pp.221–226. Wildlife Research Report No. 11. U.S. Fish & Wildlife Service, Washington.

Kampp, K. (1982). *Den Kortnaebbede lomvie Uria lomvia i Gronland—vandringer, mortalitet og beskydning: en analyse af 35 ars ringmaerkninger*. Specialerapport til Naturvidenskabelig Kandidateksamen ved Kobenhavns Universitet, Kobenhavns.

Kazama, T. (1971). Mass destruction of *Synthliboramphus antiquus* by oil pollution of Japan Sea. *Yamashina Chorui Kenkyusho Kenyuko Hokoku*, 6: 389–398.

Kenyon, K.W. (1961). Birds of Amchitka Island, Alaska. *Auk*, 78: 305–326.

Kessel, B. (1989). *Birds of the Seward Peninsula, Alaska*. University of Alaska Press, Fairbanks, Alaska.

King, C. (1984). *Immigrant Killers*. Oxford University Press, Oxford.

Kitano, K. (1981). Recent development of the studies on the North Pacific Polar frontal zone. In *Pelagic Animals and Environments around the Subarctic Boundary in North Pacific*. S.Mishima (ed), pp. 1–10. Research Institute N. Pacific Fisheries, Hokkaido University, Special Volume.

Koelink, A.F. (1972). *Bioenergetics of Growth in the Pigeon Guillemot*. M.Sc. Thesis, University of British Columbia, Vancouver.

Kondratiev, A.J. (1991). Status of the seabirds nesting in Northeast U.S.S.R. In *Seabird Status and Conservation: a Supplement*. Croxall, J.P. (ed.), pp. 159–173. I.C.B.P. Technical Publication No. 11, Cambridge.
Kostenko, V.A., Ler, P.A., Nechaev, V.A. and Shibaev, Yu.V. (1989). *Rare Vertebrates of the Soviet Far East and their Protection*. Nauka Publishing House, Leningrad.
Kozlova, E.V. (1957). *The Fauna of the USSR: Birds, Vol II(3). Charadriiformes, suborder Alcidae*. USSR Academy of Science, Moscow.
Krascheninnikov, S.P. (1786). *A Description of the Land of Kamchatka*, Vol 1, Parts 1–2. Academy of Sciences, St. Petersburg [In Russian].
Kuroda, N. (1928). Report of the biological survey of Mutsu Bay, 7 Birds of Mutsu Bay. *Scientific Report of Tohoku Imperial University, 4th Ser., Biology*, 3: 299–359.
Kuroda, N. (1954). On some osteological and anatomical characters of Japanese Alcidae (Aves). *Japanese Journal of Zoology*, 11, 311–327.
Kuroda, N. (1955). Observations on pelagic birds of the northwest Pacific. *Condor*, 57: 290–300.
Kuroda, N. (1963a). A survey of the seabirds of Teuri Island, Hokkaido, with notes on land birds. *Miscellaneous Reports of the Yamashina Institute*, 3: 363–383.
Kuroda, N. (1963). A winter seabird census between Tokyo and Kushiro, Hokkaido. *Miscellaneous Reports of the Yamashina Institute*, 3: 227–238.
Kuroda, N. (1967). Morpho-anatomical analysis of parallel evolution between Diving-Petrel and Ancient Auk, with comparative osteological data of other species. *Miscellaneous Reports of the Yamashina Institute*, 5: 111–137.
Lack, D. (1968). *Ecological Adaptations for Breeding in Birds*. Methuen, London.
Latham, J. (1785). *A General Synopsis of Birds*, vol 3, part 1. Leigh and Sotheby, London.
Lebreton, J.D. and Clobert, J. (1986). *User's Manual for Program SURGE*. CEPE/CNRS, Montpellier.
Lensink, C.C. (1984). The status and conservation of seabirds in Alaska. In *Status and Conservation of the World's Seabirds*. Croxall, J.P, Evans, P.G.H. and Schreiber, R.W. (eds), pp. 13–28. ICBP Technical Publication, No. 2, Cambridge.
Lensink, C., Gould, P. and Sanger, G. 1979. Population dynamics and trophic relationships of marine birds in the Gulf of Alaska. *U.S. Fish and Wildlife Service Ms. Report*, Anchorage.
Leschner, L.L. and Burrell, G. (1977). Populations and ecology of marine birds in the Semidi Islands. *U.S. Fish and Wildlife Service Ms. Report*, Anchorage.
Lewis, M.G. and Sharpe, F.A. 1987. *Birding in the San Juan Islands*. The Mountaineers, Seattle.
Lillard, C. (1989). *The Ghostland People*. Sono Nis Press, Victoria.
Litvinenko, N.M. and Shibaev, Yu.V. (1987). The Ancient Murrelet—*Synthliboramphus antiquus* (Gm.): reproductive biology and raising of young. In *Rasprostranenie i biologiya morskikh ptits Dal'nego Vostoka* [Distribution and biology of seabirds of the Far East], Litvinenko, N.M. (ed.), pp. 72–84. Far Eastern Science Centre of the USSR Academy of Sciences, Vladivostok.
Litvinenko, N. and Shibaev, Yu.V. (1991). Status and conservation of seabirds nesting in South-east U.S.S.R. In *Seabird Status and Conservation: a Supplement*. Croxall, J.P. (ed.), pp.175–204. ICBP Technical Publication No. 11, Cambridge.
Livezey, B.C. (1988). Morphometrics of flightlessness in the Alcidae. *Auk*, 105: 681–698.
Lobkov, E.G. (1986). *The Breeding Birds of Kamchatka*. Far East Research Centre of the USSR Academy of Science, Vladivostok.
Lockley, R.M. (1942). *Shearwaters*. J.M.Dent, London.
Macoun, J. and Macoun, J.M. (1909). *Catalogue of Canadian Birds*. Government Printing Bureau, Ottawa.

Manuwal, D.A. (1972). *The population ecology of Cassin's Auklet on Southeast Farallon Island, California*. Ph.D. Thesis, University of California, Los Angeles.

Manuwal, D.A. (1974). Effects of teritoriality on breeding in a population of Cassin's Auklet. *Ecology*, 55: 1399–1406.

Manuwal, D.A. (1984). Alcids-Dovekie, Murres, Guillemots, Murrelets, Auklets and Puffins. In *Seabirds of Eastern North Pacific and Arctic Waters*, Haley, D (ed.), pp. 168–187. Pacific Search Press, Seattle.

Martin, P.W. and Myres, M.T. (1969). Observations on the distribution and migration of some seabirds off the outer coasts of British Columbia and Washington State. *Syesis*, 1: 241–256.

Martins, R.P. (1981). *Report on a Birding Expedition to Japan*. Unpubl. Ms. in Edward Grey Institute Library, Oxford.

McTaggart-Cowan, I. (1989). Birds and mammals on the Queen Charlotte Islands. In *The Outer Shores*, Scudder, G.G.E. and Gessler, N (eds), pp. 175–186. Queen Charlotte Islands Museum Press, Skidegate, B.C.

Meyer de Schauensee, R. (1984). *The Birds of China*. Oxford University Press, Oxford.

Min, B.-Y. and Won, P.-O. (1976). An offshore winter sea bird survey on the south coast of the Korean Peninsula Mogpo Wando Yeosu area. *Yamashina Chorui Kenkyusho Kenkyu Hokoku*, 8: 53–67.

Moe, R.A. and Day, R.H. (1977). Populations and ecology of seabirds on the Koniuji group, Shumagin islands, Alaska. *U.S. Fish and Wildlife Service Unpublished Report*, Anchorage.

Munro, J.A. and Cowan, I.McT. (1947). *A Review of the Bird Fauna of British Columbia*. British Columbia Provincial Museum Special Publication No. 2, Victoria.

Munyer, E.A. (1965). Inland wanderings of the Ancient Murrelet. *Wilson Bulletin*, 77: 235–242.

Murata, E. (1958). The breeding of the Ancient Auk—*Synthliboramphus antiquus* (Gm.) on the island of Teuri, Hokkaido. *Tori*, Vol 14: 22–26.

Murie, O.J. (1959). Fauna of the Aleutian Islands and Alaska Peninsula. *North American Fauna* No. 61. US Fish and Wildlife Service, Washington.

Murray, K.G., Winnett-Murray, K. and Hunt, G.L., Jr. (1980). Egg neglect in Xantus' Murrelet. *Proceedings of the Colonial Waterbirds Group*, 3: 186–195.

Murray, K.G., Winnett-Murray, K., Eppley, Z.A., Hunt, G.L., Jr. and Schwartz, D.B. (1983). Breeding biology of the Xantus' Murrelet. *Condor*, 85: 12–21.

Murray, P. (1988). *The Vagabond Fleet*. Sono Nis Press, Victoria.

Murton, R.K. and Westwood, N.J. (1977). *Avian Breeding Cycles*. Oxford University Press, Oxford.

Nazarov, Yu.N. and Shibaev, Yu.V. (1987). Crested Murrelet, *Synthliboramphus wumizusume*, nesting on the northwestern shore of the sea of Japan. In *Rasprostranenie i biologiya morskikh ptits dalnego vostoka* [Distribution and Biology of Seabirds of the Far East], Litvinenko, M.N. (ed.), pp. 87–88. Institute of Biology and Soil Sciences of the USSR Academy of Sciences' Far Eastern Branch, Vladivostok.

Nazarov, Yu.N. and Trukhin, A.M. (1985). On the biology of *Falco peregrinus* and *Bubo bubo* on islands of the Great Peter's Bay (South Primoriye). In *Rare and Endangered Birds of the Far East*, Litvinenko, N.M. (ed.). pp. 70–76. Far East Science Centre of USSR Academy of Sciences, Vladivostok.

Nechaev, V.A. (1986). New data about seabirds on Sakhalin Island. In *Morskie Ptitsy Dalnego Vostoka* [Seabirds of the Far East], Litvinenko, N.M. (ed.), pp. 71–81. Far East Science Centre, USSR Academy of Sciences, Institute of Biology and Soil Sciences, Vladivostok.

Neff, D.J. (1956). Birds of Yang-do, Korea. *Auk*, 73: 551–555.

Nei, N.H., Hull, C.H., Jenkins, J.G., Steinbrenner, K and Bent, D.H. (1975). *Statistical Package for the Social Sciences*. 2nd ed. McGraw-Hill, New York.

Nelson, E.W. (1883). Birds of Bering Sea and the Arctic Ocean. In *Arctic Cruise of the Revenue Steamer Corwin 1881: Notes and Observations*. Government Printing Office, Washington.

Nelson, E.W. (1887). *Report upon Natural History Collections Made in Alaska between the Years 1877 and 1881*. Government Printing Office, Washington.

Nelson, J.W., Nysewander, D.R., Trapp, J.L. and Sowls, A.L. (1987). Breeding bird populations on St. Lazaria Island, Alaska. *Murrelet*, 68, 1–11.

Nelson, R.W. (1977). *Behavioural ecology of coastal peregrines (Falco peregrinus pealei)*. Ph.D. Thesis, University of Calgary.

Nelson, R.W. (1988). Do large natural broods increase natural mortality of parent peregrine falcons? In *Peregrine Falcon Populations: their Management and Recovery*, Cade, T.J., Enderson, J.H., Thelander, C.G. and White, C.M. (eds), pp. 719–728. The Peregrine Fund, Inc., Boise, Idaho.

Nelson, R.W. (1990). Status of the Peregrine Falcon, *Falco peregrinus pealei*, on Langara Island, Queen Charlotte Islands, British Columbia, 1968–1989. *Canadian Field-Naturalist*, 104: 193–199.

Nelson, R.W. and Myres, M.T. (1976). Declines in populations of Peregrine Falcons and their seabird prey at Langara Island, British Columbia. *Condor*, 78: 281–293.

Newton, I. (1979). *Population Ecology of Raptors*. T. and A.D. Poyser.

Nice, M.M. (1962). Development of behaviour in precocial birds. *Transactions of the Linnean Society of New York*, 8: 1–211.

Nol, E and Blokpoel, H. (1983). Incubation period of Ring-billed Gulls and the egg immersion technique. *Wilson Bulletin*, 95: 283–286.

Northcote, T.G., Peden, A.E. and Reimchen, T.E. (1989). Fishes of the coastal marine, riverine and lacustrine waters of the Queen Charlotte Islands. In Scudder, G.G.E. and Gessler, N. (eds) *The Outer Shores*, pp. 147–174. Queen Charlotte Islands Museum Press, Skidegate, B.C.

Nysewander, D.R, Forsell, D.J., Baird, P.A., Sheilds, D.J., Weiler, G.J. and Kogan, J.H. (1982). Marine bird and mammal survey of the eastern Aleutian islands, summers of 1980–81. *U.S. Fish and Wildlife Service Unpublished Report*, Anchorage, Alaska.

Oberholser, H.C. (1899). *Proceedings of the Academy of Natural Sciences, Philadelphia*, 51: 201.

Ogilvie-Grant, W.R. (1898). *Catalogue of the Birds in the British Museum*, vol 26. Trustees of the British Museum, London.

Ohlendorf, H.M., Bartonek, J.C., Divoky, G.J., Klaas, E.E. and Krynitsky, A.J. (1982). Organochlorine residues in eggs of Alaskan seabirds. *U.S. Fish and Wildlife Service Special Scientific Report—Wildlife* No. 245, Washington, D.C.

Osgood, W.H. (1901). *Natural History of the Queen Charlotte Islands, British Columbia*. North American Fauna, No. 21. Government Printing Office, Washington, D.C.

Pallas, P. (1811). *Zoographia Rosso-asiatica, Vol 2*. Academy of Sciences, St. Petersburg.

Peakall, D.B., Noble, D.G., Elliot, J.E., Somers, J.D. and Erickson, G. (1990). Environmental contaminants in Canadian Peregrine Falcons, *Falco peregrinus*: a toxicological assessment. *Canadian Field-Naturalist*, 104: 244–254.

Pearse, T. (1968). *Birds of the Early Explorers in the Northern Pacific*. Published by the author, Comox, B.C.

Pennant, T. (1784). *Arctic Zoology, II, class 2, Birds*. Henry Hughs, London.

Pennycuik, C.J., Croxall, J.P. and Prince, P.A. (1984). Scaling of foraging radius and growth rate in petrels and albatrosses (Procellariiformes). *Ornis Scandinavica*, 15: 145–154.

Perry, R.I., Dilke, B.R. and Parsons, T.R. (1983). Tidal mixing and summer plankton distribution in Hecate Strait, British Columbia. *Canadian Journal of Fisheries and Aquatic Science*, 40: 871–887.

Pethick, D. (1980). *The Nootka Connection*. Douglas and MacIntyre, Vancouver.
Pojar, J. and Broadhead, J. (1984). The green mantle. In *Islands at the Edge*, pp. 49–71. Douglas and MacIntyre, Vancouver.
Porter, J.M. and Sealy, S.G. (1981). Dynamics of seabird multispecies feeding flocks: chronology of flocking in Barkley Sound, British Columbia, in 1979. *Colonial Waterbirds*, 4: 104–113.
Prange, H.D. and Schmidt-Nielsen, K. (1970). The metabolic cost of swimming in ducks. *Journal of Experimental Biology*, 53: 763–777.
Pritchard, A.L. (1934). Was the introduction of the muskrat to Graham Island, Queen Charlotte Islands, unwise? *Canadian Field-Naturalist*, 48, 103.
Promislow, D.E.L. and Harvey, P.H. (1990). Living fast and dying young: a comparative analysis of life-history variation among mammals. *Journal of Zoology, London*, 220: 417–437.
Rahn, H., Paganelli, C.V. and Ar, A. (1975). Relation of avian egg weight to body weight. *Auk*, 92: 750–765.
Ratcliffe, D. (1980). *The Peregrine Falcon*. Buteo Books, Vermillion, South Dakota.
Ricklefs, R.E. (1979). Adaptation, constraint and compromise in avian post-natal development. *Biological Reviews*, 54: 269–290.
Ricklefs, R. E. (1983). Some considerations of the foraging energetics of pelagic seabirds. *Studies in Avian Biology*, 8: 84–94.
Ridgeway, R. (1896). *A Manual of North American Birds*, 2nd ed. J.B.Lippincott Co., Philadelphia.
Riedman, M.L. and Estes, J.A. (1987). A review of the history, distribution and foraging ecology of sea otters. In *Community Ecology of Sea Otters*, Vanblaricom, G.R. and Estes, J.A. (eds), pp. 4–21. Springer-Verlag, New York.
Roberson, D. (1980). *Rare Birds of the West Coast of North America*. Woodcock Publ., Pacific Grove, California.
Roby, D.D. and Brink, K.L. (1986). Breeding biology of Least Auklets on the Pribilof Islands, Alaska. *Condor*, 88: 336–346.
Rodway, M.S. (1988). British Columbia seabird inventory: report #3—census of Glaucous-winged Gulls, Pelagic Cormorants, Black Oystercatchers and Pigeon Guillemots in the Queen Charlotte Islands, 1986. *Canadian Wildlife Service Technical Report Series* No. 43. Canadian Wildlife Service, Pacific and Yukon Region, Delta, British Columbia.
Rodway, M.S. (1991). Status and conservation of breeding seabirds in British Columbia. In *Seabird Status and Conservation: a Supplement*. Croxall, J.P. (ed), pp. 43–102. ICBP Technical Publication No. 11, Cambridge.
Rodway, M.S., Hillis, N. and Langley, L. (1983). Nesting population of Ancient Murrelets on Langara Island, British Columbia. *Canadian Wildlife Service Ms. Report*, Vancouver.
Rodway, M.S., Lemon, M. and Kaiser, G.W. (1988). Canadian Wildlife Service seabird inventory report #1: East coast of Moresby Island. *Canadian Wildlife Service Technical Report Series* No 50. Canadian Wildlife Service, Pacific and Yukon Region, Delta, British Columbia.
Rodway, M.S., Lemon, M. and Kaiser, G.W. (1990). Canadian Wildlife Service seabird inventory report #2: West coast of Moresby Island. *Canadian Wildlife Service Technical Report Series* No. 65. Canadian Wildlife Service, Pacific and Yukon Region, Delta, British Columbia.
Ryder, J.P. (1980). The influence of age on the breeding biology of colonial nesting seabirds. In *Behaviour of Marine Animals, Vol 4: Marine Birds*, Burger, J. Olla, B.L. and Winn, H.E. (eds), pp. 153–168. Plenum Press, New York.
Sanger, G.A. (1972). Checklist of bird observations from the eastern North Pacific Ocean, 1955–1967. *Murrelet*, 53: 14–21.

Sanger, G.A. (1986). Diets and food web relationships of seabirds in the Gulf of Alaska and adjacent marine regions. In *Environmental Assessment of the Alaskan Continental Shelf. Final Reports of the Principal Investigators*, Vol. 45, pp. 631–771. NOAA, Anchorage, Alaska.

Sanger, G.A. (1987). Trophic levels and trophic relationships of seabirds in the Gulf of Alaska. In *Seabirds; Feeding Ecology and Role in Marine Ecosystems*. Croxall, J.P. (ed.), pp. 229–257. Cambridge University Press, Cambridge.

Sealy, S.G. (1972). *Adaptive differences in breeding biology in the marine bird family Alcidae*. Ph.D. Thesis, University of Michigan.

Sealy, S.G. (1973a). Interspecific feeding assemblages of marine birds off British Columbia. *Auk*, 90: 796–802.

Sealy, S.G. (1973b). The adaptive significance of post-hatching developmental patterns and growth rates in the Alcidae. *Ornis Scandinavica*, 4: 113–121.

Sealy, S.G. (1975a). Egg size of murrelets. *Condor*, 77: 500–501.

Sealy, S.G. (1975b). Feeding ecology of the Ancient and Marbled Murrelets near Langara Island, British Columbia. *Canadian Journal of Zoology*, 53: 418–433.

Sealy, S.G. (1976). Biology of nesting Ancient Murrelets. *Condor*, 78: 294–306.

Sealy, S.G. and Campbell, R.W. (1979). Post-hatching movements of young Ancient Murrelets. *Western Birds*, 10: 25–30.

Sealy, S.G., Fay, F.H., Bedard, J. and Udvardy, M.D.F. (1971). New records and zoogeographical notes on the birds of St Lawrence Island, Bering Sea. *Condor*, 73: 322–336.

Seebohm, H. (1890). *The Birds of the Japanese Empire*. R.H.Porter, London.

Shibaev, Yu.V. (1978). The descent to water and transition to sea life of Ancient Murrelet *Synthliboramphus antiquus* (Gm.) nestlings. In *Ecology and Zoology of Some Vertebrates of the Far East*, pp. 79–85. Institute of Biology and Soil Sciences of the USSR Academy of Sciences' Far Eastern Branch, Vladivostok.

Shibaev, Yu.V. (1987). Census of bird colonies and survey of certain bird species in Peter the Great Bay. In *Rasprostranenie i biologiya morskikh ptits dalnego vostoka* [Distribution and Biology of Seabirds of the Far East], Litvinenko, N.M. (ed.), pp. 43–59. Institute of Biology and Soil Sciences of the USSR Academy of Sciences' Far Eastern Branch, Vladivostok.

Shuntov, V.P. (1986). Seabirds of the sea of Okhotsk. In *Morskie Ptitsy Dalnego Vostoka* [Seabirds of the Far East], Litvinenko, N.M. (ed.), pp. 6–20. Institute of Biology and Soil Sciences of the USSR Academy of Science, Far East Science Centre, Vladivostok.

Sibley, C.G. and Ahlquist, J.E. (1990). *Phyllogeny and Classification of Birds*. Yale University Press, New Haven and London.

Slipp, J.W. (1942). Cassin Auklet and Ancient Murrelet on Puget Sound. *Murrelet*, 23: 18–19.

Smith, W. (1966). A second record of Ancient Murrelet from Nevada. *Condor*, 68: 511–512.

Snow, H.J. (1897). *Notes on the Kuril Islands*. John Murray, London.

Sowls, A.L., Hatch, S.A. and Lensink, C.J. (1978). *Catalog of Alaskan Seabird Colonies*. U.S. Fish and Wildlife Service, Anchorage, Alaska.

Speich, S. and Manuwal, D.A. (1974). Gular pouch development and population structure of Cassin's Auklet. *Auk*, 91: 291–306.

Speich, S.M. and Wahl, T.R. (1986). Rates of occurrence of dead birds in Washington's inland marine waters, 1978 and 1979. *Murrelet*, 67: 51–59.

Speich, S.M. and Wahl, T.R. (1989). *Catalogue of Washington Seabird Colonies*. U.S. Fish and Wildlife Service Biological Report 88(6), Washington.

Spindler, M.A. (1976). Pelagic and nearshore seabird densities in the Western Aleutian Islands as determined by transect counts in 1975 and 1976. *Aleutian Islands National Wildlife Refuge Ms. Report*, Adak, Alaska.

Springer, A.M., Kondratiev, A.Y., Ogi, H., Shibaev, Y. and Van Vliet, G.B. (in press). Status, ecology and conservation of *Synthliboramphus* murrelets and auklets. In *Proceedings of a Symposium on the Status of Seabirds in the North Pacific, February 1990*, K. Vermeer (ed.). Canadian Wildlife Service, Ottawa.

Stearns, S. (1976). Life-history tactics: a review of the ideas. *Quarterly Revue of Biology*, 51: 3–47.

Stejneger, L.H. (1885). *Results of Ornithological Explorations in the Commander Islands and Kamtschatka*. Bulletin No. 29 of the U.S. National Museum. Govt Printing Office, Washington.

Stejneger, L.H. (1887). Contributions to the natural history of the Commander Islands. *Proceedings of the U.S. National Museum*, 1887, 117–145.

Stejnegar, L.H. (1936). *George Wilhelm Steller*. Harvard University Press, Boston.

Stenzel, L.E., Page, G.W., Carter, H.R. and Ainley, D.G. (1988). *Seabird Mortality in California as Witnessed through 14 Years of Beached Bird Census*. Ms report, Point Reyes Bird Observatory, Stinson Beach, California.

Storer, R.W. (1945). The systematic position of the murrelet genus *Endomychura*. *Condor*, 47: 154–160.

Strauch, J.G., Jr. (1985). The phyllogeny of the Alcidae. *Auk*, 102: 520–539.

Stresemann, E. (1949). Birds collected in North Pacific area during Captain James Cook's last voyage (1778–1780). *Ibis*, 91: 244–255.

Stresemann, E. and Stresemann, V. (1966). Die Mauser der Vogel. *Journal für Ornithologie*, 107: 1–447.

Summers, K.R. (1974). Seabirds breeding along the east coast of Moresby Island, Queen Charlotte Islands, British Columbia. *Syesis*, 7: 1–12.

Svihla, L.A. (1952). The Ancient Murrelet at Mabton, Washington. *Murrelet*, 33: 12.

Taczanowski, L. (1893). Faune ornithologique de la Siberie Orientale. *Memoires of the Academy of Science, St. Petersburg 7th series*, 39.

Takeishi, M. (1987). The mass mortality of Japanese Murrelet *Synthliboramphus wumizusume* on the Koyashima Islet in Fukuoka. *Bulletin of the Kitakyushu Museum of Natural History*, 7: 121–131.

Taverner, P.A. (1926). *Birds of Western Canada*. King's Printers, Ottawa.

Thomson, R.E. (1981). *Oceanography of the British Columbia Coast*. Canadian Special Publications of Fisheries and Aquatic Sciences No. 56. Department of Fisheries and Oceans, Ottawa.

Thomson, R.E. (1989). Physical oceanography. In *The Outer Shores*, Scudder, G.G.E. and Gessler, N. (eds), pp. 27–64. Queen Charlotte Islands Museum Press, Skidegate, B.C.

Troy, D.M. and Johnson, S.R. (1989). Marine birds. In *Outer Continental Shelf Environmental Assessment Program Final Reports of Principal Investigators* Volume 60, pp. 355–453, U.S. Minerals Management Service, Anchorage.

Tschanz, B. (1968). Trottellummen (*Uria aalge aalge* Pont.). *Zeitschrift fur Tierpsychologie*, 4: Beiheft, 1–103.

Tso-Hsin, C. (1987). *A Synopsis of the Avifauna of China*. Science Press, Beijing.

Turner, L.M. (1886). *Contributions to the Natural History of Alaska*. Government Printing Office, Washington.

Udvardy, M.D.F. (1963). Zoogeographical study of the Pacific Alcidae. *Proceedings of the Pacific Science Congress*, 10: 85–111.

Unitt, P. (1984). *The Birds of San Diego County*. Memoire 13 of the San Diego Society of Natural History, San Diego.

Van Rossem, A.J. (1926). The Craveri Murrelet in California. *Condor*, 28: 80–83.

Verbeek, N.A.M. (1966). Wanderings of the Ancient Murrelet: some additional comments. *Condor*, 68: 510–511.

Verheyen, R. (1958). Contribution a la systematique des Alciformes. *Bulletin Institute Royal des Science Naturelles Belgique*, 34 No. 45.

Vermeer, K. (1984). The diet and food consumption of nestling Cassin's Auklets during summer, and a comparison with other plankton-feeding alcids. *Murrelet*, 65: 65–77.

Vermeer, K. and Cullen, L. (1979). Growth comparisons of a plankton and a fish-feeding alcid. *Murrelet*, 63: 34–39.

Vermeer, K. and Lemon, M. (1986). Nesting habits and habitats of Ancient Murrelets and Cassin's Auklets in the Queen Charlotte Islands, British Columbia. *Murrelet*, 67: 33–44.

Vermeer, K. and Rankin, L. (1984). Pelagic seabird populations in Hecate Strait and Queen Charlotte Sound: comparison with the west coast of the Queen Charlotte Islands. *Canadian Technical Report of Hydrography and Ocean Sciences*, No. 52.

Vermeer, K. and Rankin, L. (1985). Pelagic seabird population in Dixon Entrance. *Canadian Technical Report of Hydrography and Ocean Sciences* No. 65.

Vermeer, K., Cullen, L. and Porter, M. (1979). A provisional explanation of the reproductive failure of Tufted Puffins *Lunda cirrhata* on Triangle Island, British Columbia. *Ibis*, 121: 348–354.

Vermeer, K., Sealy, S.G., Lemon, M. and Rodway, M. (1984). Predation and potential environmental perturbances on Ancient Murrelets nesting in British Columbia. In *Status and Conservation of the World's Seabirds*, Croxall, J.P., Evans, P.G.H. and Schreiber, R.W. (eds), pp. 757–770. ICBP Technical Publication No. 2, Cambridge.

Vermeer, K., Fulton, J.D. and Sealy, S.G. (1985). Differential use of zooplankton prey by Ancient Murrelets and Cassin's Auklets in the Queen Charlotte Islands. *Journal of Plankton Research*, 7: 443–459.

Vermeer, K., Hay, R. and Rankin, L. (1987). Pelagic seabird populations off southwestern Vancouver Island. *Canadian Technical Report of Hydrography and Ocean Science*, No. 87.

Vyatkin, P.S. (1986). Nesting colonies of colonial birds in the Kamchatka region. *Morskie Ptitsy Dalnego Vostoka* [Seabirds of the Far East]. Litvinenko, N.M. (ed.), pp. 20–36. Far East Science Centre, USSR Academy of Science, Institute of Biology and Soil Sciences, Vladivostok.

Wahl, T.R. (1975). Seabirds in Washington's offshore zone. *Western Birds*, 6: 117–134.

Wahl, T.R. (1978). Seabirds in the northwestern Pacific Ocean and south central, Bering Sea in June 1975. *Western Birds*, 9: 45–66.

Wahl, T.R., Speich, S.M., Manuwal, D.A., Hirsch, K.V. and Miller, C. (1981). *Marine Bird Populations of the Strait of Juan de Fuca, Strait of Georgia, and Adjacent Waters in 1978 and 1979*. EPA-600/F-81-156, U.S. Environmental Protection Agency, Washington.

Webb, R.L. (1988). *On the Northwest: Commercial Whaling in the Pacific Northwest, 1790–1967*. University of British Columbia Press, Vancouver.

Wehle, D.H.S.(1983). The food, feeding and development of young Tufted and Horned Puffins in Alaska. *Condor*, 85: 427–442.

Weimerskirch, H., Bartle, J.A., Jouvetin, P. and Stahl, J-C. (1988). Foraging ranges and partitioning of feeding zones in three species of southern Albatrosses. *Condor*, 90: 214–219.

White, C.M., Emison, W.B. and Williamson, F.S.L. (1973). DDE in a resident Aleutian Island peregrine population. *Condor*, 75: 306–311.

Whitely, H. (1867). Notes on birds collected near Hakodadi in Northern Japan. *Ibis* N.S., 3: 193–211.

Wilbur, S.R. (1987). *The Birds of Baja California*. University of California Press, Berkeley.

Willett, G. (1914). Birds of Sitka and vicinity, southeastern Alaska. *Condor*, 16: 71–91.

Willett, G. (1915). Summer birds of Forrester Island, Alaska. *Auk*, 32: 295–305.

Willett, G. (1920). Comments upon the safety of seabirds and upon the "probable" occurrence of the northern Bald Eagle in California. *Condor*, 22: 204–205.

Williams, G.C. (1966). Natural selection, the cost of reproduction, and a refinement of Lack's principle. *American Naturalist*, 100: 687–690.

Xantus de Vesey. (1860). *Proceedings of the Academy of Natural Sciences, Philadelphia* (1859), 11: 299.

Yamashina, Y. (1961). *Birds in Japan; a Field Guide*. Tokyo News Service, Tokyo.

Ydenberg, R.C. (1989). Growth-mortality trade-offs and the evolution of juvenile life-histories in the Alcidae. *Ecology*, 70: 1494–1506.

Young, C.J. (1927). A visit to the Queen Charlotte Islands. *Auk*, 44: 38–43.

Zammuto, R.M. (1986). Life histories of birds: clutch size, longevity and body mass among North American game birds. *Canadian Journal of Zoology*, 64: 2739–2749.

Zwiefelhofer, D.C. and Forsell, D.J. (1989). *Marine birds and mammals wintering in selected bays of Kodiak Island, Alaska: a five-year study*. Ms. report, U.S. Fish and Wildlife Service, Anchorage.

Index

Acanthomysis, 54
Active Pass, 63
Adak Island, 43, 59
Agglomerate island, 209
Akutan Pass, 18
Alaska, 15, 17, 25–30, 67, 94
Alaska Peninsula, 18, 20, 29
Alca antiqua, 9, 12
Alcidae, 7, 12
Alder, 74
Aleutian Islands, 20, 25–29, 37, 44, 58, 63, 131, 169, 221
Aleuts, 17
Alexander Archipelago, 69
Amagat Island, 29
Amchitka Island, 43
Ammodytes, 52–54
Amphipod, 54
Anchorage, Alaska, 68
Andreanof Islands, 27
Arichika Island, 36
Audubon, 12, 15
Avatcha, 10

Baja California, 41, 48, 59
Bald Eagle, 74, 82, 91, 93
Barkley Sound, 61
Barnacle, 83
Barren Islands, 29
Beaufort Sea, 73
Beaver, 80, 89
Bering, 10, 17
Bering Island, 17, 25, 68
Bearing Sea, 25, 37, 50, 59, 68, 79
Bill morphology, 55
Biro Island, 39
Bischof Islands, 36
Black-tailed Gull, 58
Black-throated Guillemot, 12
Boulder Island, 36
Brachyramphus, 12, 14, 216
Brandt's Cormorant, 93
Breeding habitat, 37–41
Bristol Bay, Alaska, 63
Britain, 67
British Columbia, 9, 15, 20, 50, 53, 59, 64, 67, 79, 82
Brood patch, 99–100, 132–133
Brown Creeper, 92
Brunnich's Guillemot, 3
Buldir Island, 26, 35, 38, 54, 172

California, 35, 48, 50, 59–61, 63–64, 68, 72, 175
California Condor, 222
California Current, 72
Calypso Orchid, 76
Canada Goose, 37
Cape St. James, 69, 73, 82
Capelin, 54
Capella Cove, 87
Captain Cook, 10, 68
Carroll Island, Washington, 37
Cassin's Castle, 91–92, 218
Cassin's Auklet, 14, 58, 74, 78, 87, 92, 157–158, 216
Castle Rock, 29
Cepphini, 14
Cepphus, 12, 14, 172, 212, 218
Chenlusan Island, 22
Chestnut-backed Chickadee, 91
Chibaldo Island, 22
Chika Islets, 18
China, 20, 38, 82
Chiton, 83
Chocolate Lily, 78
Chukchi Sea, 62
Coal Harbour, Vancouver I., 79
Coats Island, N.W.T., 179
Commander Islands, 9, 11, 15, 17, 25, 37, 43, 55, 59, 68
Common Eider, 158
Common Murre, 3, 215
Cook Inlet, 29
Copper Island (Commander Is.), 25
Craveri's Murrelet, 8, 13–14, 41, 48, 56, 219
Crested Murrelet, 13
Cumshewa Inlet, 69, 72, 93
Cut-throat Trout, 74

Daikoku-Jima, 24
Dall's Porpoise, 73
Danjo Islands, 39
Dawson Caribou, 80
De-Kastri Bay, 24
Decapod, 54
Deer Mouse, 80, 83, 89, 123, 207, 217
Departure of chicks, 18
Dixon Entrance, 30, 34, 69
Dodge Point, Lyell Island, 82, 207, 209
Double-crested Cormorant, 93
Dovekie, 17, 217

East Siberia, 15
El Nino, 72
Elephant Seal, 73
Elymus/Calamagrostis, 38
Endomychura, 13–14
Ermine, 68, 80, 89
Euphausia, 52–54, 74
Euphausid, 52–56, 58, 74, 92, 94, 192, 216
Europe, 67

Far Eastern Marine Sanctuary, 22
Farallon Islands, California, 72, 218
Farid Ud-din Attar, 1
Feeding behaviour, 57–58
Fin Whale, 94
First World War, 79
Flicker, 91
Fork-tailed Storm Petrel, 18, 92
Forrester Island, 16, 18, 29–30, 34, 44, 122, 125
Fox Sparrow, 91
Fox Islands, 26–27
Frederick Island, 16, 30, 34, 37, 44, 157, 205
Fukuoka Prefecture, 39

Gammerid, 56
Gathering Ground, 44, 142–153
George Dixon, 68
Glaucous-winged Gull, 57–58, 93–94
Gmelin, 12, 15
Golden-crowned Kinglet, 91
Goose Islands, 50
Graham Island, 16, 30, 54, 69, 71–72, 74, 79–80, 151
Great Auk, 7, 213
Great Lakes, 61
Grey Whale, 73
Grizzly Bear, 29
Gulf of Alaska, 29, 37, 44, 54, 59, 62, 71
Gulf of St. Lawrence, 222
Gwaii Haanas, 69, 80

Haida, 17, 78, 80, 89
Hairy Woodpecker, 91
Halibut, 74
Hamgyong Pukto, 22
Hanaguri Island, 39
Harbour Seal, 73, 123
Haro Strait, 63
Hashira Island, 39
Hecate Strait, 4, 8, 37, 50, 61, 63, 69, 71–73, 80, 94, 129, 151, 192–193, 196, 217, 223
Helgeson Island, 30
Hermit Thrush, 91
Heron, 1
Herring Gull, 94
Herring, 54, 58, 74

Hippa Island, 16, 30, 37, 44, 151
Hokkaido, 24
Hong Kong, 59
Honshu, 24, 39, 59
Hoopoe, 1, 224
House Island, 35, 163, 209
Huckleberry, 76, 81
Humpback Whale, 73, 79, 94, 192

Indian Pipes, 76
Islands of the Four Mountains, 27
Izu Peninsula, 39

Japan, 9, 15, 18, 38, 50–51, 59–61, 63–64, 67–68
Japan Wild Bird Society, 39
Japanese Murrelet, 12–14, 39–41, 56, 218
Jiangsu Province, 22
Juan Perez Sound, 209

Kamchadals, 10
Kamchatka, 10–11, 15, 24, 37, 44
Kanagawa Prefecture, 39
Kangwon do, 22
Kanmuri umizusume, 13
Karamzin Island, 22
Kelp, 72, 78, 126
Killer Whale, 73
Kingsway Rock, 93
Kittlitz's Murrelet, 12–13
Klykov Island, 22
Kodiak Island, 29, 59, 63
Koniuji Island, 27
Korea, 15, 22, 38, 57, 61
Koyashima Islet, 39
Krascheninnicof, 9
Kuril(e) Islands, 11, 15, 24, 36–37, 39, 59
Kuroshio Current, 39
Kyushu, 39

Langara Island, 16, 30–35, 43–44, 49, 52, 57, 64, 81, 118, 129, 131, 151, 158, 169, 177, 214, 221
Laskeek Bay, 72, 87, 93
Laskeek Bay Conservation Society, 224
Leach's Storm Petrel, 92
Least Auklet, 179
Limestone Islands, 36, 43, 81, 143–146, 199, 202, 208–209
Limpet, 83, 91
Little Diomede Island, 62
Little Brown Bat, 89
Lodgepole Pine, 74, 87
Low Island, 36, 73, 78, 91–93, 109, 143
Lucy Island, 16, 72
Lupin, 78
Lyell Island, 43, 163

Macrocystis, 72
Magadan, 24
Marbled Murrelet, 8, 10, 13–14, 78
Marine habitat, 62–64
Massett Inlet, 72
Measurements, 101–103
Mergulus antiquus 12, 15
Middleton Island, 29
Mikomoto Island, 39
Mimiana Island, 39
Mink, 68, 80
Mixed species flocks, 57–58
Moneron Island, 24
Monkey Flower, 78
Monterey, 41, 59, 63
Moore Islands, 37, 50
Moresby Island, 16, 30, 69, 74, 199
Moult, 8
Murchison Island, 35
Muskrat, 80
Mussel, 83
Mutsu Bay, Honshu, 59

Naden Harbour, 79
Nan-do, 22
Navarin Basin, 59
New Zealand, 91
Newport, Oregon, 62
Niigata Prefecture, 59
Ninstints, 78
Nishi Island, 22
Non-breeding range, 59–62
Nootka Sound, 68
Northern Fur Seal, 68, 73
Northwestern Crow, 91

Oil, 4
Okinoshima Island, 39
Oldsquaw, 72
Orange-crowned Warbler, 91
Oregon, 63

Pacific Loon, 94
Pacific Ocean, 20, 67, 94, 123
Pacific White-sided Dolphin, 73
Pallas, 9–11, 14–15, 17
Panama Canal, 67
Parrakeet Auklet, 10
Pavlof Islands, 29
Peale's Peregrine Falcon, 18, 34–36, 54, 82, 91
Pelagic Cormorant, 92–93
Penguin, 7
Pennant, 9, 11, 14–15
Penzhinskaya Gulf, 11, 20, 24
Peromyscus, 80
Pesticides, 34–35

Peter the Great Bay, 22–24, 37–39, 44, 54, 58
Pigeon Guillemot, 10, 72
Pine Marten, 80, 89
Pine Siskin, 91
Pleistocene, 37, 84
Plumage, 7–9
Point Conception, 41, 61
Polychaete, 54
Potentilla, 78
Predation
  By foxes, 27–29
  By cats, 41
  By raccoons, 36
  By rats, 24, 35, 39
  By bears, 29
Pribilof Islands, 26
Primoriye, 39
Prince William Sound, 62
Puffin, 10
Puget Sound, 61

Qingdao, 22
Queen Charlotte Sound, 50, 69, 72, 79, 193, 208
Queen Charlotte City, 69, 81
Queen Charlotte Islands, 2, 13, 15–16, 20, 30–37, 43–44, 58, 61–62, 64, 67, 89, 123, 125, 142, 155, 162, 169, 172, 206, 208, 221, 223

Raccoon, 81
Ramsay Island, 163, 209
Rankine Island, 125, 163
Rat Islands, 26
Raven, 82, 91
Razorbill, 213, 215, 217
Red Crossbill, 91
Red Columbine, 78
Red-breasted Sapsucker, 91
Red-necked Phalarope, 94
Reef Island, 8, 35, 43–44, 49, 57–58, 62–63, 69, 73, 76, 80, 83, 85–210
Rennell Sound, 72
Rhinoceros Auklet, 14, 22, 57–58, 72, 78, 94, 152, 158
River Otter, 68, 80, 89
Rockfish, 54
Rocky Mountain Elk, 81
Rocky Mountains, 67
Rose Spit, 72
Rose Harbour, 79
Rufous Hummingbird, 91
Ruskii Island, 22
Russia, 67

Sable, 68
Sakhalin, 24, 39

Index

Salal, 76, 81
Salmon, 74
Salmonberry, 76
Sambondake Island, 39
San Benito Island, 41
San Diego, 59, 63
San Francisco, 41
Sanak Island, 18, 27, 38
Sandlance, 52–54
Sandman Reefs, 20, 29
Sandspit, 16, 69, 72, 85, 89, 193
Sangam-Jima, 24
Santa Barbara Island, 41
Saw-whet Owl, 91, 123, 207
Sea urchin, 72, 78
Sea Pigeon Island, 36
Sea of Japan, 24, 54, 59
Sea of Cortez, 41
Sea of Okhotsk, 11, 24
Sea Otter, 68, 72, 78–79
Semidi Islands, 29
Shandong, 22
Shanghai, 22
Shantar Islands, 24
Sharp-shinned Hawk, 91
Shelikof Strait, 29
Shikotan Island, 44
Ship Rat, 81–83
Shrew, 80, 89
Shumagin Islands, 10, 29, 54, 57
Siberia, 20
Single Delight, 76
Sitka Black-tailed Deer, 80–81, 89
Sitka Spruce, 22, 74, 78–79, 87
Skedans Islands, 36, 78, 91
Skedans, 78
Skidegate, 81
Skidegate Inlet, 69, 73
Skincuttle Inlet, 36, 163
Skinkana (Haida), 13
Skunk Cabbage, 76
Smallpox, 79
Sockeye Salmon, 22
Song Sparrow, 91–92
Sooty Shearwater, 57–58, 94, 192
South Moresby, 16, 30, 36, 49–50, 69, 72, 80–81, 89, 207, 223
South Low Island, 78, 143
Spain, 67
Sperm Whale, 79
Spitz Island, 29
Squirrel, 89
St. Elias Head, Alaska, 11
St. Lawrence Island, 26, 50, 59
St. Lazaria Island, 29
Stable isotopes, 54
Staging area, see Gathering ground
Starischkow (Starichkov) Island, 10, 24, 38

Steelhead, 74
Steller, 9–11, 14, 17
Steller's Jay, 11, 91
Steller's Sea Cow, 68
Steller's Sealion, 73, 89
Storm Petrel, 9, 29, 35
Strait of Georgia, 61, 63
Strait of Juan de Fuca, 61, 63
Strait of Korea, 59
Subspecies, 9
Swainson's Thrush, 91
Systematics, 13–14

Taiwan, 59
Takung Tao, 22
Talan Island, 24
Talunkwin Island, 79
Tanu, 78
Taxonomy, 9–13
Temminck's Auk (Guillemot), 12–13
Terrestrial environments, 74–78
Teuri Island, 24
Thick-billed Murre, 3, 179, 183, 189, 215
Thurston Harbour, 80
Thysanoessa, 52–53
Timing of laying, 44–48
Toad, 80, 89
Tobishima, 24
Tokyo, 39, 61
Townsend's Warbler, 91–92
Triangle Island, 72, 152
Tsushima Current, 39
Tufted Puffin, 22, 82, 213

U.S. Fish and Wildlife Service, 20, 25
Unalaska, 11, 59
Uria, 217
Uria senicula, 9
Uria antiqua, 12
Ussuriland, 61

Vancouver, 68
Vancouver Island, 37, 50, 54, 61, 68, 73
Varied Thrush, 91
Verkhovskii Island, 22, 125
Victoria, B.C., 50, 52, 57, 63, 79
Vladivostok, 22
Vocalizations, 104–107, 134–136

Walleye Pollock, 54
Washington State, 15–16, 37, 61, 63
Weasel, 68
Western Europe, 67
Western Hemlock, 74, 87, 158
Western Red cedar, 74, 81, 87
Westport, Washington, 61
Whiskered Auklet, 14

Windy Bay, Lyell Island, 76
Winter Wren, 91, 140

Xantus' Murrelet, 13–14, 22, 41, 48, 56, 126, 170, 175, 179, 183, 207, 217–218

Yakutat Bay, 62
Yang-do, 22
Yellow Cedar, 76
Yellow Sea, 20
Yew, 76
Yokohama Bay, 39